Uncertainty and Operations Research

Decision analysis based on uncertain data is natural in many real-world applications, and sometimes such an analysis is inevitable. In the past years, researchers have proposed many efficient operations research models and methods, which have been widely applied to real-life problems, such as finance, management, manufacturing, supply chain, transportation, among others. This book series aims to provide a global forum for advancing the analysis, understanding, development, and practice of uncertainty theory and operations research for solving economic, engineering, management, and social problems.

Bhaskar Basu · Jitendra Kumar Jha ·
Indrajit Mukherjee · Raghu Nandan Sengupta
Editors

Data Analytics for Decision Making towards Business Excellence

Through the Lens of Retail and Financial Services

Springer

Editors
Bhaskar Basu
Information Systems Area
Xavier Institute of Management, XIM University
Bhubaneswar, India

Jitendra Kumar Jha
Department of Industrial and Systems Engineering
Indian Institute of Technology Kharagpur
Kharagpur, India

Indrajit Mukherjee
Shailesh J. Mehta School of Management
Indian Institute of Technology Bombay
Mumbai, India

Raghu Nandan Sengupta
Department of Industrial and Management Engineering
Indian Institute of Technology Kanpur
Kanpur, Uttar Pradesh, India

ISSN 2195-996X ISSN 2195-9978 (electronic)
Uncertainty and Operations Research
ISBN 978-981-95-2428-0 ISBN 978-981-95-2429-7 (eBook)
https://doi.org/10.1007/978-981-95-2429-7

This Springer imprint is published by the registered company Springer Nature Singapore Pte Ltd.
The registered company address is: 152 Beach Road, #21-01/04 Gateway East, Singapore 189721, Singapore

Preface

In today's hyper-dynamic marketplace, where data flows faster than decisions are often made, the ability to interpret, integrate, and act upon analytical insights has become not only a differentiator but a necessity. This volume, *Data Analytics for Decision Making Towards Business Excellence Through the Lens of Retail and Financial Services*, was conceived with the intent of capturing this shift and exploring how organizations across sectors are embracing data science to drive strategic and operational superiority.

Retail and financial services—two sectors central to global economic and social activity—stand at the confluence of massive data generation and the urgent need for agile decision-making. These industries serve millions daily, and their digital footprints offer a treasure trove of behavioral, transactional, and environmental information. Yet, harnessing this data intelligently is no trivial feat. This book aims to showcase the transformative potential of analytics in these domains by curating a diverse and interdisciplinary set of chapters that illuminate both academic rigor and industry relevance.

The editorial team, comprising scholars from premier institutions: XIM University, IIT Kharagpur, IIT Bombay, and IIT Kanpur, have come together with a shared vision: to provide a compendium of actionable research, methodological innovations, and real-world applications. Our collaboration seeks to bridge the academic and practitioner gap, offering readers conceptual clarity alongside empirical grounding.

The book is structured in two parts—Part I: The Retail Lens and Part II: The Financial Services Lens—each examining the distinctive characteristics, challenges, and opportunities of data-driven decision-making in its respective domain. In the retail section, readers will explore themes such as inventory optimization, customer engagement, omni-channel strategies, and the emerging role of digitalization in crafting personalized experiences. From pricing effectiveness to service recovery mechanisms, this section provides both conceptual frameworks and case-based insights that can inform managerial actions. In the financial services section, the focus shifts to areas including behavioral finance, algorithmic trading, risk modeling, sentiment analysis, and forecasting. Chapters delve into how advanced techniques such as deep learning, econometrics, and big data analytics are reshaping financial markets and institutions. The inclusion of policy-making models and tools for detecting market manipulation adds further breadth to this discourse.

What sets this volume apart is its deliberate blend of theoretical depth and applied relevance. Readers—from business students to seasoned professionals—will find value in its methodological diversity, ranging from econometric modeling to machine learning applications. The book encourages a critical appreciation of not only what analytics can do, but also how decision contexts influence outcomes. It pushes the reader to think about data beyond dashboards—to consider ethics, interpretability, and sustainability. This book also stands as a testament to collaborative scholarship. The contributing authors, a mix of established researchers and emerging voices, have brought in perspectives from academia, industry, and policy. Each chapter reflects a commitment to rigor, clarity, and relevance, and we believe this multi-vocality enriches the reading experience.

Ultimately, the goal of this volume is to foster more informed and impactful decisions in retail and finance using data analytics as both lens and lever. As technological capabilities evolve and market complexities deepen, we hope this book serves as a beacon for those navigating the intersection of data and strategy.

We invite the readers to explore, reflect, and engage—and to continue shaping the future of analytics-led decision making.

Bhubaneswar, India — Bhaskar Basu
Kharagpur, India — Jitendra Kumar Jha
Mumbai, India — Indrajit Mukherjee
Kanpur, India — Raghu Nandan Sengupta

Acknowledgements

The journey of bringing this book to life has been both enriching and collaborative, and I am deeply grateful to all those who played a role along the way.

First and foremost, I extend my heartfelt thanks to my family, whose unwavering support, patience, and encouragement sustained me throughout this endeavour. Your belief in my work gave me the strength to see it through.

To my fellow editors, Raghu (R.N. Sengupta), Indro (Indrajit Mukherjee) and Jhaji (Jitendra K. Jha): thank you for your keen insights, constructive critiques, and spirited discussions. Your intellectual companionship has been vital in shaping the narrative and structure of this work.

I am sincerely thankful to each chapter contributor for your dedication and scholarship. Your perspectives, research, and writing have added immense depth and diversity to this volume.

A special word of appreciation goes to the tireless production team—your attention to detail, design sensibility, and professionalism brought coherence and elegance to the final product.

To the publishing team led by Nupoor Madam: thank you for your guidance, efficiency, and commitment to excellence. Your stewardship has helped bring this work to a wider audience with quality and clarity.

Finally, I am indebted to the senior management at XIM University for your steadfast support and belief in the academic value of this project. Your encouragement created the environment in which this book could flourish.

To all who shared in this journey, my deepest thanks.

Bhaskar Basu

Contents

Data Analytics for Decision Making Towards Business Excellence

Bhaskar Basu

The digital transformation of the retail industry has resulted in a data-intensive environment, compelling organizations to adopt advanced analytics for informed decision-making. The first section of this book examines the role of data analytics as a strategic tool to attain business excellence, with an emphasis on contemporary challenges, emerging trends, and best practices within the retail services context. Real-world case studies and everyday examples are incorporated to demonstrate practical applications and outcomes.

Retail enterprises are increasingly navigating a landscape defined by digital convergence, omnichannel complexity, and heightened consumer expectations. In this context, data analytics has emerged as a cornerstone of strategic decision-making, enabling organizations to derive actionable insights from vast and varied data sources. The shift from intuition-based to evidence-based management is not merely a trend but a necessity for achieving sustainable business excellence. The retail sector has witnessed a dramatic escalation in data generation, driven by e-commerce platforms, mobile applications, social media, and IoT-enabled devices. This transformation aligns with the five V's of Big Data—volume, velocity, variety, veracity, and value. Retailers now collect structured and unstructured data from point-of-sale systems, customer feedback loops, loyalty programs, and digital footprints, necessitating sophisticated analytical frameworks for interpretation and application.

Despite the proliferation of data, several challenges persist:

- Data Fragmentation: Disparate systems and siloed data repositories hinder the development of a unified customer view.
- Governance and Compliance: Ensuring data accuracy, ethical usage, and regulatory compliance remains a complex undertaking.
- Talent Shortage: The demand for skilled data professionals often exceeds supply, limiting the scalability of analytics initiatives.

B. Basu (✉)
Information Systems Area, Xavier Institute of Management, XIM University, Bhubaneswar, India
e-mail: bhaskar@xim.edu.in

B. Basu et al. (eds.), *Data Analytics for Decision Making towards Business Excellence*, Uncertainty and Operations Research, https://doi.org/10.1007/978-981-95-2429-7_1

- Latency in Decision-Making: The inability to process real-time data can result in missed opportunities and suboptimal outcomes.

Retailers are increasingly leveraging advanced analytics to gain a competitive advantage:

- Predictive and Prescriptive Analytics: These tools enable demand forecasting, dynamic pricing, and inventory optimization.
- Customer Journey Mapping: Analytics platforms track multichannel interactions to enhance personalization and engagement.
- AI and Automation: Chatbots and robotic process automation (RPA) streamline operations and improve customer service.
- Geospatial Intelligence: Location-based analytics inform store placement, regional marketing, and logistics.
- Sustainability Analytics: Data is used to monitor carbon footprints, optimize supply chains, and meet ESG benchmarks.

Important use cases pertaining to the application of data analytics in this sector are as follows:

- Enhancing Customer Experience: Nike's "Nike By You" initiative allows customers to design their own shoes using an online customization platform. This personalization strategy, powered by customer data and feedback, has significantly increased engagement and brand loyalty. Adidas introduced interactive mirrors in fitting rooms that recognize items brought in by customers and display product information, alternative sizes, and complementary items. This blend of physical and digital interaction enhances the in-store experience and encourages upselling.

In everyday life, customers experience this personalization when shopping online and receiving tailored product recommendations based on browsing history or previous purchases—an experience now expected across major retail platforms.

- Pricing Strategies in Action: Retailers employ a variety of pricing strategies to remain competitive and profitable.
- Dynamic Pricing**:** Amazon adjusts prices in real time based on demand, competitor pricing, and customer behavior. For instance, during festive seasons, prices for high-demand items fluctuate frequently to optimize sales and margins.
- Psychological Pricing: Brands like Bata and Big Bazaar use ₹99 or ₹199 pricing to create a perception of affordability, even though the difference from the next round number is minimal.
- Value-Based Pricing**:** Apple maintains premium pricing for its products, justified by perceived quality and brand equity. Despite higher costs, customers are willing to pay for the ecosystem and experience.
- Discount Pricing**:** Retailers like DMart and Walmart use everyday low pricing (EDLP) to attract price-sensitive customers, relying on volume sales to maintain profitability.

These strategies are increasingly data-driven. Retailers analyze historical sales, competitor pricing, and customer segments to determine optimal price points for different products and regions.

- Digitalization in Retailing

IKEA has embraced augmented reality (AR) through its app, allowing customers to visualize furniture in their homes before purchasing. This reduces product returns and enhances customer confidence. Reliance Retail in India has implemented an omnichannel strategy integrating physical stores with digital platforms like JioMart. Customers can browse online, place orders via app, and choose home delivery or in-store pickup. This seamless experience is powered by real-time inventory and customer data analytics. Starbucks uses its mobile app to offer loyalty rewards, enable mobile ordering, and personalize promotions. Customers can view their favorite drinks, earn points, and even track the origin of their coffee beans—an example of how digitalization enhances transparency and engagement.

In daily life, consumers benefit from these innovations through conveniences like scanning QR codes for product details, receiving digital receipts, or using voice assistants to add items to shopping carts.

To institutionalize analytics-driven decision-making, retailers should adopt the following practices:

- Strategic Alignment: Ensure that analytics initiatives are aligned with organizational goals and customer value propositions.
- Infrastructure Investment: Deploy scalable cloud-based platforms and real-time analytics engines.
- Cross-Functional Integration: Foster collaboration among marketing, operations, IT, and finance to maximize data utility.
- Ethical Governance: Develop transparent policies for data usage, algorithmic accountability, and consumer privacy.

The integration of data analytics into retail operations aligns with the principles of Total Quality Management (TQM) and the Business Excellence Model. By enabling continuous improvement, customer-centric innovation, and agile decision-making, analytics serves as a catalyst for long-term value creation. Moreover, the feedback mechanisms inherent in analytics systems promote organizational learning and resilience. In an era defined by digital disruption and consumer empowerment, data analytics is not merely a tool but a strategic imperative. Retailers that can effectively harness data to inform decisions, personalize experiences, and optimize operations will be well-positioned to achieve business excellence. The case studies and real-life examples presented herein underscore the transformative potential of analytics and offer a roadmap for academic inquiry and practical implementation.

Looking ahead, the future of data-driven retailing will be shaped by several key developments. AI-powered recommendation systems are evolving into foundational components of retail ecosystems, offering hyper-personalized experiences by integrating structured and unstructured data—from browsing behavior to contextual cues like time of day and location. Retailers will increasingly adopt context-aware and generative AI technologies, enabling dynamic content creation, visual search, and voice-based shopping interfaces that redefine customer interaction. Moreover, predictive delivery models and AI-enhanced supply chains will become central to customer satisfaction. These systems will anticipate demand, optimize logistics, and offer flexible delivery timelines, enhancing transparency and trust. As data privacy regulations tighten, ethical AI practices and privacy-preserving techniques like federated learning will gain prominence. Ultimately, the convergence of real-time analytics, omnichannel integration, and intelligent automation will empower retailers to move from reactive operations to proactive, customer-centric innovation. Those who invest in robust data strategies today will lead the retail landscape of tomorrow.

Part I of the book compiles contributions of authors with interests in the retail sector. The first chapter by Qehaja explores the digital transformation in the retail sector, highlighting the crucial role of data analytics in achieving competitive advantage. As the retail landscape evolves digitally, understanding how data analytics informs strategic decision-making is vital. The chapter begins with an overview of digital transformation's significance and associated challenges and opportunities, discussing trends like omnichannel integration, AI, and data privacy. A notable case study is Tesco, which has effectively harnessed data analytics to enhance its operations and customer engagement. By leveraging customer purchase data, Tesco has personalized marketing efforts, optimized inventory management, and improved supply chain efficiencies. This approach not only drives sales but also enhances customer loyalty, illustrating how data-driven strategies can lead to significant competitive advantages. The concept of competitive advantage refers to the unique edge retailers gain from effectively using digital technologies and data analytics to improve operational efficiency and deliver customer insights. Nearly 80% of European retailers, for instance, have adopted advanced analytics, with factors like data availability and technology driving this trend. Digital transformation signifies integrating digital technologies across retail businesses, fundamentally reshaping how they operate and deliver value. This transformation emphasizes agility and customer-centricity. Organizational changes towards data-driven decision-making and cross-disciplinary collaboration are also essential. Other trends affecting digital transformation include the growth of e-commerce, mobile shopping, and personalized marketing. E-commerce has notably driven retail expansion in the U.S., with 82% of consumers making online purchases recently. While online shopping is rising, traditional in-store experiences remain important. Understanding these transformations is crucial for anticipating future changes and leveraging digital technologies for successful retail strategies.

The second chapter by Shirsat *et al.* presents a comprehensive study on how a leading Indian private sector bank strategically adopted data analytics to transform customer experience and operational efficiency. Originating from a merger between a corporate and infrastructure financing bank and a retail finance leader, the combined institution capitalized on its distinct strengths to serve a broader customer base. The bank placed data analytics at the centre of its operations, creating an integrated data infrastructure and centralized data warehouse to unify customer information and support real-time decision-making. It leveraged advanced tools such as machine learning, natural language processing, and robotic process automation to personalize offerings, enhance risk management, and streamline operations. Key initiatives included personalized marketing, predictive modelling for customer engagement, fraud detection using transactional anomaly analysis, and automation of repetitive tasks. These efforts led to measurable improvements in customer satisfaction, operational cost savings, and financial performance, including higher loan origination volumes and net interest margins. Despite these successes, the bank encountered challenges such as data inconsistency, privacy concerns, and employee resistance. These were mitigated through robust data governance, cybersecurity measures like multi-factor authentication, and extensive staff training programs aimed at fostering a data-driven culture. The study's novelty lies in its focus on an Indian bank—filling a gap in the literature largely centred on advanced economies. It extends theoretical frameworks like customer journey mapping and CRM by incorporating real-time analytics in a developing market context. Practically, it offers valuable insights for banks in emerging economies, advocating for tailored customer services, data-driven risk assessment, and the adoption of scalable analytics solutions. The research is grounded in publicly available secondary sources, enriched by the author's first-hand experience in the bank's leadership during its formative years.

The next chapter by Valase and Kaur offers a thorough exploration of returns management and reverse logistics within the evolving landscape of omnichannel (OC) retailing. As retailers transition from multichannel models to fully integrated OC strategies, the seamless orchestration of online and offline channels has become essential for enhancing customer experience and operational efficiency. However, this integration also brings new challenges, especially in handling product returns, which are complex, costly, and critical to customer satisfaction. The review emphasizes that, unlike traditional or even multichannel models, OC retailing enables cross-channel returns, where customers can return items bought online to physical stores. This flexibility improves customer convenience and offers retailers opportunities for cost reduction, inventory optimization, and cross-selling. Yet, it also introduces logistical intricacies, including decisions about return routing, processing locations (e.g., store, distribution centre, or return centre), and gatekeeping—each with trade-offs between speed, cost, and sustainability. The study highlights the scarcity of literature specifically addressing reverse logistics in OC contexts, especially from a closed-loop supply chain (CLSC) perspective. It points out the lack of thematic organization within existing research and the need for updated, structured reviews to guide future studies. Key themes needing further exploration include customer behavior in return channels, return speed and its financial impact, sustainability in reprocessing returns, and strategic use of reprocessed products. Ultimately, the chapter calls for a systematic review to synthesize fragmented insights and offers a roadmap for both academic inquiry and practical implementation, laying the groundwork for more resilient, customer-centric OC retail operations. This review seeks to bridge gaps in current understanding and support retailers in navigating the rising tide of returns with thoughtful, data-driven strategies.

The fourth chapter by Wang *et al.* in this section explores the challenge of managing "demon customers" in the direct response industry—customers who either return products excessively or default on payment (bad debt). While past research has focused mostly on the implications of returns, this study brings in the underexplored dimension of bad debt, especially in contexts where credit is extended via "bill me later" or Buy Now, Pay Later (BNPL) models without formal credit checks. Drawing on a large dataset from a Spanish book retailer using direct mail marketing, the study investigates how customer behaviour related to returns and bad debt affects firm profitability. A key insight is that customers who frequently return products are also more likely to default, challenging the assumption that returners are more conscientious than defaulters. The authors test various predictive models to evaluate how historical data—such as previous orders, mailings, returns, and debt—can forecast future customer behaviour. Surprisingly, the research finds that forbidding product returns might enhance profitability under certain conditions, as the operational costs associated with processing returns can sometimes exceed the cost of lost products due to bad debts. The models demonstrate that combining return behaviour and bad debt history improves the precision of customer targeting strategies. This dual-focus approach offers important implications: it reveals a positive correlation between return and bad debt tendencies and shows how incorporating bad debt data leads to better marketing decisions. Notably, this paper is among the first to examine these issues in a segment of consumers often overlooked—those without access to credit cards. Ultimately, the research encourages firms to re-evaluate return policies and embrace more nuanced, data-driven targeting algorithms, particularly as BNPL platforms expand in the post-pandemic era. It also hints at broader strategic trade-offs between customer convenience, operational efficiency, and long-term profitability.

The final chapter in the section by Kaur *et al.* proposes an advanced deep learning framework to improve sentiment analysis of online customer reviews by addressing issues like review bias and data imbalance. Online reviews significantly influence consumer decisions, but they often exhibit extreme sentiment due to voluntary bias—either overly positive or negative. Traditional sentiment analysis models struggle with such skewed data and the semantic richness of customer language. To mitigate these challenges, the authors introduce a hybrid model combining CNN (Convolutional Neural Networks), BiLSTM (Bidirectional Long Short-Term Memory), and a topic-level attention mechanism. CNN layers capture local textual features, while BiLSTM processes sequential and contextual dependencies in both directions. The topic-based attention layer enhances the model's ability to focus on contextually relevant words, leading to more accurate sentiment classification. The study integrates word embedding techniques like Word2Vec, GloVe, and FastText to capture semantic relationships between words. It also employs oversampling techniques—SMOTE, Borderline SMOTE, and SVMSMOTE—to address class imbalance, enhancing classification performance. The model's effectiveness is validated on Amazon and Yelp review datasets, showing strong performance in terms of accuracy, precision, recall, and F1 score. Among the oversampling methods, SMOTE performs best, while Word2Vec offers a favourable trade-off between accuracy and computational efficiency. GloVe provides high precision and recall, and FastText supports faster processing where necessary. Additionally, the research leverages FastAPI to facilitate integration of the model with external applications. The study concludes with future directions such as exploring field-specific embeddings, improving model interpretability, adapting to dynamic review environments, and integrating multimodal data (text and visuals). Overall, this work contributes to the advancement of NLP by offering a robust, scalable, and interpretable sentiment analysis model tailored for real-world e-commerce applications.

In the contemporary financial services landscape, data analytics has become a strategic cornerstone of organizational agility, innovation, and regulatory compliance. The second part of the book critically explores the growing data intensity of the sector and investigates how analytics enables evidence-based decision-making aligned with business excellence principles. It examines prevailing challenges, emerging trends, best practices, and technological interventions, using case-based insights and real-life examples to illustrate the integration of big data methodologies.

The financial services sector is characterized by its complex, data-driven environment, where decision-making processes must balance risk, compliance, market responsiveness, and customer satisfaction. With the rise of fintech, digital banking, and algorithmic trading, the industry has transitioned into a high-frequency, high-stakes data ecosystem. Data analytics is no longer a supporting function but a catalyst for enterprise transformation, offering granular visibility into customer behavior, portfolio performance, and systemic vulnerabilities. The evolution from descriptive to predictive and prescriptive analytics is reshaping finance into an intelligent, responsive domain. This article outlines the sector's trajectory toward data maturity and the role of analytics in fostering business excellence—a concept rooted in continuous improvement, innovation, and stakeholder value. Financial institutions generate vast datasets through an array of digital interfaces—mobile apps, ATMs, trading platforms, customer relationship management (CRM) systems, and blockchain networks. The "5Vs" of big data—volume, velocity, variety, veracity, and value—are fully manifest in this context. Structured datasets (e.g., transactions, KYC information) coexist with unstructured sources (e.g., social media sentiment, voice recordings, and chat logs), requiring hybrid analytical approaches involving data mining, natural language processing, and real-time visualization. The imperative is not just to store or process data, but to translate it into actionable, ethically sound decisions.

Despite advancements, financial services face several entrenched challenges:

- Legacy Systems and Data Silos: Many firms operate under fragmented IT architectures that inhibit data integration and holistic insight generation.
- Regulatory Overhead: Compliance with evolving mandates—such as the GDPR, Basel Accords, and India's Digital Personal Data Protection Act—necessitates robust data lineage and audit capabilities.
- Data Governance Deficiencies: Issues related to metadata management, data stewardship, and quality control undermine analytic reliability.
- Talent Scarcity: There remains a marked deficit in professionals adept at both financial modeling and modern data science techniques, impeding cross-functional innovation.

Contemporary developments indicate a paradigm shift in how financial data is leveraged:

- Predictive Credit Risk Models: Firms like Experian and HDFC use machine learning to enhance credit scoring through behavioral and alternative data streams.
- Fraud Detection and Cyber Risk Mitigation: JPMorgan Chase and American Express deploy AI to detect anomalous transactions in near-real-time, significantly reducing exposure.
- Personalized Banking Services: Robo-advisors such as Betterment and Wealthfront dynamically adjust investment strategies based on real-time changes in client goals and market conditions.
- Sustainability Analytics: ESG (Environmental, Social, and Governance) compliance is driving the use of data to evaluate climate risk, carbon exposure, and social impact in asset allocation.
- RegTech Innovations: Automated compliance monitoring tools analyze complex legal texts and transactional patterns to alert firms to risks of non-compliance.

Important use cases pertaining to the application of data analytics in this sector are as follows:

- Walmart: Driver-Based Financial Forecasting: Although traditionally a retail giant, Walmart's financial planning transformation offers valuable insights for financial institutions. Facing inflationary pressures and supply chain volatility, Walmart adopted driver-based forecasting to link operational levers—like foot traffic and fuel prices—to financial outcomes. This enabled dynamic scenario modeling and improved forecast accuracy across business units.
- JPMorgan Chase: COiN Platform: JPMorgan's Contract Intelligence (COiN) platform uses natural language processing to review legal documents, reducing 360,000 h of manual work to seconds. This not only improved operational efficiency but also enhanced compliance accuracy and audit readiness.
- Ant Group: Inclusive Credit Scoring: Ant Group's microloan platform leverages behavioral and transactional data from its Alipay ecosystem to assess creditworthiness for individuals and small businesses lacking formal credit histories. This data-driven model has expanded financial inclusion across underserved populations in China and Southeast Asia.
- American Express: Fraud Detection and Customer Insights: American Express applies real-time analytics to detect fraudulent activity and personalize offers. By analyzing transaction patterns and customer behavior, the company has reduced fraud losses while increasing customer engagement and retention.
- SBI's YONO Platform: The State Bank of India's YONO app integrates banking, shopping, and investment services. It uses analytics to segment users, predict service needs, and cross-sell financial products. This has helped SBI modernize its customer engagement strategy and compete with private and digital-first banks.

In daily life, customers interact with data-driven financial services in numerous ways:

- Mobile Banking Apps: Users receive real-time alerts for unusual transactions, personalized loan offers, and spending insights based on categorized expenses.
- UPI and Digital Wallets: Platforms like Google Pay and PhonePe use analytics to suggest frequent payees, offer cashback promotions, and detect payment anomalies.
- Credit Card Recommendations: Based on spending behavior, banks suggest cards with better rewards or lower interest rates.
- Insurance Premiums: Telematics-based car insurance adjusts premiums based on driving behavior, tracked via mobile apps or IoT devices.
- Chatbots and Virtual Assistants: AI-powered assistants help users check balances, schedule payments, and resolve queries without human intervention.

These examples underscore how data analytics is embedded in the financial routines of millions, enhancing convenience, personalization, and security.

To institutionalize a data-driven culture, financial institutions should:

- Define a Clear Analytics Maturity Model: Map current capabilities against desired outcomes across descriptive, diagnostic, predictive, and prescriptive analytics stages.
- Adopt Scalable Infrastructures: Invest in cloud-native architectures, data lakes, and containerized analytic environments to handle high-throughput workloads securely.
- Ensure Ethical AI Governance: Build transparency, fairness, and interpretability into AI systems, especially in risk-sensitive areas like loan approvals and fraud detection.
- Advance Workforce Literacy: Upskill existing professionals through interdisciplinary training in statistics, machine learning, and domain-specific frameworks such as IFRS 9 or Solvency II.
- Integrate Human Oversight with Automation: While AI enhances efficiency, human judgment must remain integral, particularly in nuanced domains such as regulatory decisions and customer grievance resolution.

Business excellence in financial services is increasingly defined by adaptability, compliance agility, and stakeholder trust. Data analytics acts as a linchpin across these axes, facilitating continuous performance improvement via insights-driven optimization. The deployment of advanced analytics aligns with total quality and lean finance principles, reducing waste, enhancing customer satisfaction, and aligning operational goals with strategic vision. Furthermore, integrated feedback loops—powered by real-time dashboards and predictive alerts—enable proactive decision-making under volatile conditions.

The trajectory of financial services is irrevocably tied to the evolution of data analytics. The coming decade will witness further convergence of AI, quantum computing, edge analytics, and decentralized finance (DeFi) systems—each introducing novel data paradigms and risk vectors. Future-ready financial institutions will prioritize explainable AI, federated learning, and automated model governance to remain both agile and accountable. Open banking APIs and composable financial architectures will enable modular, customer-centric services that can be rapidly deployed and scaled. Real-time data streaming and semantic interoperability will allow institutions to process and act on insights with sub-second latency, transforming everything from fraud detection to investment advisory. Moreover, hyper-personalization will become the norm. Customers will expect financial services that anticipate their needs, adapt to their behaviors, and deliver seamless experiences across platforms. Institutions that can integrate behavioral analytics, contextual data, and ethical AI will lead in customer trust and loyalty.

From an academic perspective, these dynamics warrant interdisciplinary research integrating finance, data science, behavioral economics, and regulatory studies. Analytics-enabled decision-making is not a passing phase but the blueprint for systemic excellence in modern financial ecosystems. As data becomes the lifeblood of finance, the institutions that master its flow will shape the future of the industry.

Part II of the book compiles contributions of authors with interests in the financial services sector. The first chapter in this section by Gandhudi *et al.* explores an innovative approach to stock price prediction by merging textual data analysis with cutting-edge quantum machine learning (QML) techniques. Recognizing the powerful influence of news headlines on market sentiment, the authors extract features using Bag of Words, TF-IDF, and N-gram models to capture semantic nuances from a rich dataset of global news and corresponding stock movements. Classical machine learning methods—such as SVM, Random Forest, and deep learning models like LSTM and GRU—have long been used for financial forecasting but face challenges like overfitting, high computational cost, and difficulty modelling non-linear relationships. To address these issues, the study introduces a suite of QML models—including Quantum Logistic Regression, Quantum SVM, Quantum KNN, and Quantum Random Forest—leveraging principles like superposition and entanglement for enhanced processing of high-dimensional data. Among the models tested, Quantum Random Forest achieved the best performance, with 96% accuracy and 97% precision, outperforming both other quantum models and conventional algorithms. The data pre-processing pipeline included rigorous steps like stop word removal, stemming, lemmatization, and normalization to ensure clean input for training. The chapter emphasizes that accurate stock prediction can improve investor decision-making, risk management, and strategic planning. The integration of quantum algorithms enables faster and more efficient exploration of financial data, making them well-suited for the dynamic, complex nature of stock markets. By bridging the gap between classical and quantum computing, this work positions QML as a transformative tool in financial analytics. It also outlines future directions, including refining quantum models, expanding datasets, and incorporating more contextual information like multi-modal data, all aimed at building robust and real-time predictive systems in finance. The chapter thus contributes to both academic discourse and practical strategies in the era of quantum-enhanced data science.

The next chapter by Vinitha and Saravanan investigates the relationship between Environmental, Social, and Governance (ESG) scores and the volatility of mutual fund returns, with a focus on sustainable investing in the post-COVID-19 landscape.

Analyzing data from 64 mutual funds rated by Morningstar in April 2023, the study employs cross-sectional regression to assess whether higher ESG scores influence return volatility over a one-year period. Findings indicate that mutual funds linked to companies with high social and governance risk scores—as well as stronger sustainability ratings—exhibit higher volatility in returns. This suggests that while these funds may carry greater short-term risk, they also hold potential for higher long-term returns. The research emphasizes that ESG considerations are not just ethical but also strategic, attracting investors who prioritize sustainable practices. It also notes a surge in ESG-related investments in India, where the mutual fund industry has seen a sixfold increase in assets under management over a decade. Despite resilience during global disruptions, challenges remain: ESG data is often inconsistent, disclosure standards vary, and larger firms may disproportionately benefit from current scoring systems. The study aligns with prior global research highlighting that socially responsible investment (SRI) portfolios can show greater resilience during financial crises. However, it also acknowledges limitations such as smaller sample sizes and biases in sustainability scores. Importantly, the paper calls for greater clarity on material ESG disclosures tailored to industry-specific and regional contexts. Ultimately, the document encourages investors, regulators, and mutual fund companies to view ESG not just as a compliance metric, but as a framework for long-term value creation, risk mitigation, and societal impact. It urges companies to align profits with ethical responsibilities, warning that ignoring ESG imperatives—especially in the face of climate risk and global scrutiny—places firms on precarious ground.

The third chapter in this section by Sharma *et al.* examines the impact of the January 2023 Hindenburg Research report on the stock prices and financial performance of the Adani Group, one of India's largest conglomerates. Employing both event study methodology and panel data regression, the study quantifies abnormal returns and tracks financial indicators before and after the report's release. It finds that the report caused significant negative abnormal returns across Adani Group companies and adversely affected quarterly financial performance, highlighting strong market sensitivity to allegations of fraud and stock manipulation. The Hindenburg report accused the Adani Group of inflated stock valuations, over-leverage, promoter stock pledging, and shell company misuse. Despite Adani's firm rebuttal, the market reaction was immediate and severe, wiping billions in market capitalization and triggering downgrades, lawsuit filings, and global investor scrutiny. The study dissects company-wise cumulative abnormal returns over three distinct post-report phases (0–33, 34–66, and 67–100 days), identifying both short-term panic and long-term volatility. Drawing on behavioural finance, the research notes how investor sentiment, media influence, and weak regulatory mechanisms contributed to heightened volatility and sustained damage to reputation and trust. It argues that the Indian market, while reactive, displayed semi-strong form inefficiency, as seen in slow information absorption and herd-like retail investor behaviour. Beyond financial metrics, the study emphasizes broader implications, calling for corporate governance reforms, improved disclosure norms, audit independence, and regulatory vigilance (particularly by SEBI). It also highlights the socioeconomic impact, including job insecurity and loss of wealth among retail investors. The authors position the Adani-Hindenburg episode as a wake-up call for emerging markets like India to bolster oversight, market integrity, and investor protection. It adds valuable insight into the ripple effects of corporate scandals on financial ecosystems.

The final chapter in the section by Biswas *et al.* explores the intricate relationship between Environmental, Social, and Governance (ESG) performance and financial efficiency in energy firms across developing countries from 2016 to 2023. Using data from the Bloomberg Terminal, the researchers apply Data Envelopment Analysis (DEA) and the Malmquist Productivity Index to assess energy firms' operational and productivity efficiency, linking ESG inputs—like environmental compliance, stakeholder engagement, and governance quality—to output indicators such as Return on Equity, Return on Assets, and Return on Capital. The findings reveal substantial year-to-year efficiency fluctuations, largely influenced by ESG integration, technological advancement, and regulatory dynamics. Firms specializing in solar energy and petroleum, especially those based in China and Malaysia, consistently demonstrate high or perfect efficiency. This reflects their proactive adoption of sustainable practices, innovations in photovoltaic technologies, and favourable policy environments. The study also contextualizes the ESG-financial performance nexus within global sustainability trends, emphasizing how ESG integration enhances risk management, credit ratings, and investor confidence, especially in resource-intensive sectors like energy. It draws upon stakeholder theory and the resource-based view (RBV) to illustrate how ESG strategies create intangible assets, improve compliance, and reinforce long-term resilience. Importantly, the research identifies critical gaps in ESG implementation in developing nations, highlighting disparities in regulatory frameworks, disclosure standards, and corporate social responsibility maturity. It underscores the role of technological innovation and global capital flows in bridging the sustainability divide, especially amid challenges like COVID-19, fluctuating energy demand, and climate risk. Overall, the chapter advocates for more nuanced, context-sensitive ESG strategies tailored to the developmental and industrial realities of emerging markets. By aligning ESG initiatives with economic development goals, energy firms in developing countries can achieve both financial competitiveness and sustainable impact, fostering resilience in a rapidly evolving global energy landscape.

Hopefully, the compilation of chapters will provide a glimpse of the enormous possibilities of future research in data analytics through the lens of retail and financial services.

The Retail Lens

Digital Transformation in Retail: Leveraging Data Analytics for Competitive Advantage

Albana Berisha Qehaja

Abstract This chapter comprehensively examines the digital transformation in the retail sector, emphasizing the crucial role of data analytics in achieving competitive advantage. As retailers face unprecedented challenges and opportunities in an increasingly digital marketplace, understanding how data analytics can inform strategic decision-making is vital. The chapter follows a structured approach, beginning with an introduction to digital transformation in retail and its significance in today's business landscape. It addresses the challenges and opportunities associated with digital transformation initiatives in the retail sector and explores emerging trends such as omnichannel integration, artificial intelligence, and data privacy considerations. Case studies from leading retailers offer practical illustrations of data-analytics-driven digital transformation strategies. This chapter contributes to the scholarly discourse on digital transformation and data analytics in the retail sector, offering valuable insights for academics, researchers, and industry professionals interested in leveraging data analytics for a competitive advantage in retail settings. The digital transformation of the retail sector is not merely an academic topic but a real-world phenomenon reshaping our economy and society. By understanding these changes, we can better anticipate the future and harness the power of digital technologies and data analytics to create more efficient, customer-centric, and successful retail businesses.

Keywords Digital transformation · Retail sector · Data analytics · Competitive advantage · Data privacy

1 Introduction

The digital revolution has significantly affected various industries, including retail. This chapter provides a comprehensive examination of digital transformation within the retail sector, focusing on the crucial role of data analytics in gaining a competitive

A. Berisha Qehaja (✉)
University of Prishtina, Prishtina, Republic of Kosovo
e-mail: albana.berisha@uni-pr.edu

B. Basu et al. (eds.), *Data Analytics for Decision Making towards Business Excellence*, Uncertainty and Operations Research, https://doi.org/10.1007/978-981-95-2429-7_2

advantage (Akter et al., 2022; Bresciani et al., 2021; Dahiya et al., 2022; Garg & Goyal, 2019; Hänninen et al., 2018; Khanom, 2023; Kumar, 2024; Leão & da Silva, 2021; Ramazanov et al., 2021; Rossato & Castellani, 2020; Shah, 2022).

Digital transformation is the "process that aims to improve an entity by triggering significant changes to its properties through combinations of information, computing, communication, and connectivity technologies" (Vial, 2019, p. 118). In retail, this transformation involves integrating digital technology into all areas of a business, fundamentally changing how retailers operate and deliver value to customers (Hänninen et al., 2018; Rossato & Castellani, 2020).

As part of this transformation, the use of big data has become increasingly important. Big data refers to large volumes of data generated at high velocity and variety, which traditional data processing methods cannot handle efficiently; unlike traditional data analytics, which deals with more minor, structured datasets, big data analytics can process and analyze vast, diverse datasets, uncovering hidden patterns and insights that drive more informed decision-making. This shift enables retailers to gain deeper insights and make more informed decisions (Dekimpe, 2020; Ferreira et al., 2020).

In recent times, there has been a significant increase in scholarly articles on business, management, and information systems that highlight the influence of digital technologies (Schneider & Kokshagina, 2021). Alexandrova and Kochieva (2021) argue that the evolution of digital technologies in retail networks is transforming the retail landscape. Technologies such as artificial intelligence (AI), blockchain, and cloud computing are poised to shape the future of retail continually.

Globalization, digital technologies, and multinational trading corporations are altering purchasing and consumption patterns, influencing the sphere of circulation (Ramazanov et al., 2021). Retailers are increasingly recognizing the need to adapt to the digital age, with digital transformation becoming a strategic imperative for successful retailers (Quinones et al., 2023). As retailers face challenges and opportunities in an increasingly digital marketplace, understanding how to leverage digital technologies becomes vital (Akter et al., 2022).

Data analytics leverages data, statistical algorithms, and machine-learning techniques to uncover significant patterns and insights (Dahiya et al., 2022), which is crucial in driving digital transformation initiatives. In retail, data analytics can inform strategic decision-making, anticipate market trends, and enhance customer experiences, thereby driving competitive advantage (Hänninen et al., 2018; Khanom, 2023; Kumar, 2024). Big data analytics can help organizations achieve sustainable competitive advantage by providing firm-specific knowledge that is valuable, rare, inimitable, and non-substitutable (Hagiu & Wright, 2020).

In the context of digital transformation and data analytics, competitive advantage refers to the unique advantage a retailer gains over its competitors through the effective use of digital technologies and data analytics (Leão & da Silva, 2021; Ramazanov et al., 2021). This advantage can take various forms, such as improved operational efficiency, enhanced customer insights, and the ability to predict future trends (Pinto et al., 2023).

Digital transformation in retail has become a global phenomenon. Kothari and Seetharaman (2020) investigate its influence on the UAE's retail banking industry, providing insights applicable to the retail sector. According to a recent study by Coresight Research (2019), the retail industry prioritizes the adoption of analytics. Nearly 80% of European retailers have already implemented advanced data analytics capabilities. Factors such as data availability, technology, and competition continue to drive further adoption of analytics in the retail sector.

In the following sections, we delve deeper into these topics, providing a structured approach to understanding digital transformation in retail and the role of data analytics in driving competitive advantage. We will explore real-world examples, discuss advanced data analytics techniques and tools, examine challenges and opportunities, and investigate future trends in the industry. By understanding these changes, we can better anticipate the future and harness the power of digital technologies and data analytics to create more efficient, customer-centric, and successful retail businesses.

2 Digital Transformation in Retail

Digital transformation has significantly changed various industries in recent decades. Newcomers challenge established firms by utilizing digital technology to build platform ecosystems and drive industry-wide changes (He et al., 2023). In the retail sector, emerging technologies are revolutionizing operations. Machine learning and big data analytics can help retailers process the vast amounts of customer data they have gathered, while robots and automation systems are going from factories to warehouses and distribution centers. The Internet of Things (IoT) allows for precise tracking of products, whether on store shelves or across the globe, offering retailers the opportunity to integrate technology and data into the physical retail environment (McKinsey & Company, 2020).

In the retail industry, digital transformation signifies integrating digital technology into every facet of a retail business, fundamentally changing how retailers function and deliver value to their customers. This transformation is not solely about moving traditional retail online but also about reshaping retail businesses to be agile, flexible, and customer-centric (Bresciani et al., 2021). Factors driving the adoption of retail analytics include enhancing the shopping experience, improving operational efficiency, reducing costs, increasing profitability, and boosting competitiveness (Coresight Research, 2019).

Digital transformation also implies changes in organizational structures and cultures, such as transitioning from management-driven to data-driven decision-making, promoting cross-disciplinary collaboration instead of isolated silos, and embracing agile, experimental, and improvisational approaches over risk-averse strategies (Levallet & Chan, 2018). A 2017 global survey on digital transformation revealed that only 25% of organizations had successfully transitioned into digital businesses, with 41% in the transformation process and 34% primarily discussing the trend rather than implementing it (Kane, 2017).

Over the years, several trends have influenced digital transformation efforts in the retail sector. These include the growing prominence of e-commerce, mobile shopping, personalized marketing, and AI and machine learning integration into retail operations (Akter et al., 2022). E-commerce has been a significant driver of retail growth in the United States, accounting for 40% of the expansion since 2016. A survey by McKinsey revealed that 82% of US consumers had made online purchases in the past three months, with 42% of millennials expressing a preference for online shopping over in-store shopping (McKinsey & Company, 2020). Despite the rise of online shopping, the traditional in-store experience remains relevant.

Analyzing historical sales data and market trends allows retailers to optimize inventory levels, reducing overstock and stockouts and minimizing storage costs. For instance, a retailer can use predictive analytics to forecast demand for different products, ensuring they stock the right amount of each item (Araz et al., 2020; Leão & da Silva, 2021).

By analyzing customer behavior data, retailers can select the most effective marketing channels, enhancing customer engagement and improving conversion rates. For example, a retailer might discover that social media ads are more effective than email campaigns for a particular product. They can achieve better results by reallocating their marketing budget to focus more on social media (Barua, 2018; Dahiya et al., 2022).

Omnichannel retailing integrates various shopping channels to provide a seamless customer experience. Data analytics is crucial in understanding customer preferences and behavior across different channels, allowing retailers to offer personalized experiences and improve overall customer satisfaction. For instance, a retailer can track a customer's journey from browsing products online to purchasing in-store and use this data to provide personalized recommendations and promotions (Cakir et al., 2021; Proskurnina et al., 2021).

Stephenson (2018) discusses how businesses can enhance decision-making and competitive positioning through the strategic use of big data. Businesses can understand customer behaviors, optimize operations, and anticipate future market trends by leveraging data analytics. Reinartz et al. (2019) explore how digital transformation affects the retail value chain, leading to more efficient operations and improved customer experiences.

By leveraging data analytics, retailers can segment their customers according to purchasing behaviors, preferences, and demographics, enabling them to provide tailored products, services, and experiences. Furthermore, demographic information, such as age, gender, income level, and location, can be used for segmentation. Insights obtained, such as younger customers favoring online shopping and older customers preferring physical stores, can guide retailers' marketing strategies and service offerings.

Barua (2018) highlights how European retail firms can gain a competitive advantage by using a digital platform for the segmentation process, suggesting that digital platforms provide sophisticated tools for customer segmentation and personalization.

Looking ahead, Pantano (2020) suggests that technological advancements, changes in consumer behavior, and the evolving competitive landscape will influence the future of digital transformation in retail. Zaki (2019) argues that digital transformation will bring about the next generation of services in the retail sector, offering new ways for retailers to engage with customers and deliver value.

Cakir et al. (2021) presented a case study on the digital transformation of a medium-sized retailer through omnichannel retailing, offering valuable insights into the practical application of digital transformation strategies in the retail industry. Chernova et al. (2020) discussed how digital transformation can enhance customer value, increase customer loyalty, and increase sales. Dahiya et al. (2022) discussed the strategic role of firm-specific knowledge in gaining a competitive advantage through big data analytics, emphasizing how retailers can leverage their unique knowledge and data effectively.

The impact of digital transformation on retail business is significant. This leads to changes in consumer behavior and market dynamics. Consumers expect seamless experiences across all channels, personalized interactions, and instant gratification (Kothari & Seetharaman, 2020; Chorna et al., 2021). This has forced retailers to reassess their strategies and operations, intensifying competition in the market dynamics, with retailers constantly innovating to remain competitive (Leão & da Silva, 2021). Digital transformation has also opened up new opportunities for retailers, such as reaching a global audience, creating highly targeted marketing campaigns, and gathering and analyzing vast customer data. Khanom (2023) emphasizes implementing strategies prioritizing customers, assimilating emerging technologies, and formulating agile and adaptable business models to maintain relevance in the digital transformation era.

Digital transformation is reshaping the retail industry, driven by technological changes, consumer expectations, and market dynamics. Retailers who leverage digital technologies and data analytics gain a significant competitive advantage. The future of retail lies in the successful integration of digital technologies and data analytics, transforming the retail landscape and redefining the shopping experience for consumers worldwide.

3 Advanced Data Analytics Techniques and Tools

To fully leverage the potential of data analytics in retail, it is essential to understand the advanced techniques and tools available. These tools can significantly enhance decision-making processes and operational efficiency:

Machine Learning Algorithms: Machine learning, a subset of artificial intelligence, allows systems to autonomously learn and enhance their performance based on past data and experiences without requiring explicit instructions for every task. In retail, machine learning algorithms can predict customer behavior, optimize inventory management, and personalize marketing strategies. Standard algorithms include decision trees, support vector machines, and neural networks (Akter et al., 2022; Lutfi et al., 2023; Marr, 2021).

Artificial Intelligence: AI encompasses various technologies, including natural language processing and computer vision. These technologies can analyze customer sentiment from social media, monitor in-store customer behavior, and enhance customer service through chatbots. Tools like IBM Watson and Google AI provide powerful solutions for retail analytics (Akter et al., 2022; Marr, 2021; Pantano, 2020).

Big Data Platforms: Apache Hadoop and Apache Spark are designed to handle large-scale data processing. These tools enable retailers to efficiently store, process, and analyze vast data. Managing the massive datasets generated by modern retail operations is critical to gaining insights and driving business strategies (Abbu & Gopalakrishna, 2022; Aversa et al., 2021; Dahiya et al., 2022; Dekimpe, 2020; Ferreira et al., 2020; Khan et al., 2022; Shah, 2022).

Business Intelligence (BI) Tools: BI tools like Tableau, Power BI, and Qlik Sense offer robust data visualization and reporting capabilities. These tools help retailers convert complex data into actionable insights through interactive dashboards and reports, facilitating informed decision-making (Aversa et al., 2021; Bejta, 2021; Dahiya et al., 2022; Fan & Geerts, 2022; Shah, 2022).

Predictive Analytics Software: Software such as SAS Predictive Analytics and RapidMiner enables retailers to forecast future trends by analyzing historical data. Predictive analytics helps anticipate customer demand, optimize pricing strategies, and reduce supply chain risks (Araz et al., 2020; Julie, 2015; Khanom, 2023; Pinto et al., 2023; Shah, 2022).

Customer Relationship Management (CRM) Systems: CRM such as Salesforce and Microsoft Dynamics integrate customer data from various touchpoints to provide a unified view of customer interactions. This comprehensive view helps retailers personalize customer experiences, enhance customer service, and improve loyalty (Bresciani et al., 2021; Levallet & Chan, 2018; Quinones et al., 2023).

Data Warehousing Solutions: Amazon Redshift and Google BigQuery offer scalable data warehousing capabilities. These platforms allow retailers to consolidate data from multiple sources, facilitating comprehensive analysis and reporting essential for strategic planning and performance management (Aversa et al., 2021; Bresciani et al., 2021; Chernova et al., 2020; Ferreira et al., 2020; Ramazanov et al., 2021).

Category Management: Involves managing product categories as strategic business units, streamlining inventory, and optimizing product placement. Data analytics is crucial in category management by helping retailers understand customer preferences, sales patterns, and inventory levels. For example, in e-commerce, data analytics can analyze customer browsing and purchasing behavior to identify the most popular categories and the products with the highest demand. This information can optimize product listings, improve inventory management, and enhance customer experience (Abbu & Gopalakrishna, 2022; Bresciani et al., 2021).

In brick-and-mortar retail, data analytics can monitor sales data, customer traffic patterns, and inventory turnover rates to identify which product categories are performing well and which need adjustments. These data enable retailers to make strategic decisions about product placement, promotional strategies, and inventory replenishment (Cote, 2021; Torres, 2022).

Trend Analysis: Data analytics is crucial in identifying and forecasting retail trends. Retailers can uncover emerging trends and better understand customer preferences by examining extensive datasets from various sources, including sales records, social media platforms, and market analyses. For example, retailers can use predictive analytics to forecast future demand for specific products based on historical sales data and current market trends. This helps retailers stock the appropriate products exactly when they are in demand, reducing the risk of overstock or stockouts (Araz et al., 2020; Barua, 2018).

Data analytics can also help retailers track the effectiveness of marketing campaigns and promotions, providing insights into what resonates with customers and what does not. Regularly analyzing and tracking data allows retailers to stay ahead of market trends and strategically adjust to maintain a competitive advantage (Dahiya et al., 2022; Proskurnina et al., 2021).

Incorporating these practices enhances operational efficiency, helps retailers meet customer expectations, and improves overall business performance. By leveraging data analytics for category management and trend analysis, retailers can make data-driven decisions that drive growth and profitability (Garg & Goyal, 2019; Hagiu & Wright, 2020; Leão & da Silva, 2021; Pinto et al., 2023; Reinartz et al., 2019; Rossato & Castellani, 2020).

4 Leveraging Data Analytics for Competitive Advantage in Retail

Today's retail industry can be broadly classified based on the channels through which retailers offer their products and services. There are three main categories:

- **Bricks-and-clicks**: These are retailers that operate both online and offline.
- **Online retailers**: These retailers operate exclusively through online channels.
- **Brick-and-mortar**: These retailers have a physical presence and operate solely through offline channels (Barua, 2018).

An increasing number of traditional brick-and-mortar retailers are transitioning to a brick-and-clicks model. In this model, they use online channels for consumer outreach, product promotion, and advertising, even if not for direct sales.

Data analytics is gaining increasing prominence in the retail sector. It involves examining, cleaning, transforming, and modeling data to uncover valuable information, infer conclusions, and support decision-making (Dahiya et al., 2022). This empowers retailers to deeply understand their customers, optimize operations, and boost sales.

Insights gathered from data analytics can provide retailers with a wealth of information about customer behavior, preferences, and purchasing patterns, enabling them to tailor their strategies effectively.

Data analytics can help retailers improve their supply chain management, inventory control, and marketing strategies. By analyzing sales data, retailers can anticipate demand for various products, optimize inventory, and minimize storage costs. Furthermore, data analytics can give retailers a competitive advantage by enabling them to make data-driven decisions, predict market trends, and enhance customer experiences, ultimately gaining a competitive advantage (Hänninen et al., 2018; Khanom, 2023; Rossato & Castellani, 2020).

Data analytics techniques are commonly used in retail settings to extract meaningful insights from large volumes of data. These techniques can be broadly categorized into four types (Insights Desk, 2023).

As explained by Cote (2021), they are:

- Descriptive analytics, which addresses the question, "*What happened*?"
- Diagnostic analytics, which responds to the question, "*Why did this happen*?"
- Predictive analytics, which answers the question, "*What might happen in the future*?"
- Prescriptive analytics, which guides on, "*What should we do next*?"

When utilized in tandem, each form of data analytics can contribute to specific objectives, providing a holistic view of data that is instrumental in shaping an organization's strategy and decision-making processes (Cote, 2021).

4.1 Descriptive Analytics

Descriptive analytics, the analysis of historical data to understand past events, is crucial for retailers as it provides insights into their past performance and helps identify trends. It is the most widely used form of analytics in retail, enabling businesses to gain a deeper understanding of their customer base, sales trends, inventory levels, and more. Descriptive analytics finds applications in various aspects of the retail sector:

a. *Understanding Sales Trends*: Retailers can use descriptive analytics to understand sales trends over time by analyzing historical sales data to identify patterns and inform future business decisions (Cote, 2021; Torres, 2022).
b. *Analyzing Customer Behavior*: Descriptive analytics helps retailers understand their customer bases by analyzing data on customer behavior, such as purchase history and browsing behavior, to gain insight into customers' preferences and shopping habits (Akter et al., 2022).
c. *Inventory Management*: Retailers utilize descriptive analytics for effective inventory management. They analyze data on inventory levels and sales to maintain the right amount of stock and avoid overstock or stock-out situations.

Additionally, according to Malenkov et al. (2021), digital transformation significantly impacts retail chain stores, particularly in areas such as inventory management. Digital technologies allow for more efficient and effective inventory management, leading to improved operational performance and competitive advantage (Akter et al., 2022; Pinto et al., 2023; Ramazanov et al., 2021; Rossato & Castellani, 2020).

The integration of omnichannel strategies into the digital transformation process can further enhance inventory management by providing a seamless shopping experience across all channels. This allows retailers to leverage data analytics to gain a holistic view of their inventory, meet customer demands, and improve overall business performance (Cakir et al., 2021; Proskurnina et al., 2021; Quinones et al., 2023).

By harnessing big data, retailers can gain deeper insights into their inventory, allowing for more accurate forecasting and strategic decision-making (Dahiya et al., 2022; Stephenson, 2018). This ultimately leads to enhanced operational efficiency and a more competitive position in the market (Khan et al., 2022; Pinto et al., 2023; Ramazanov et al., 2021; Rossato & Castellani, 2020).

d. *Understanding Market Conditions*: Descriptive analytics helps retailers understand market conditions by analyzing data on market trends and competitor activity. This allows them to gain insights into the market environment and adjust their strategies accordingly (Dahiya et al., 2022).

Walmart uses descriptive analytics to understand sales trends and manage their inventory effectively. By analyzing sales data, Walmart can forecast demand, streamline replenishment processes, and reduce stockouts or overstock situations (Torres, 2022).

4.2 Diagnostic Analytics

Diagnostic analytics involves using data to identify the reasons behind observed trends and relationships among variables. It is often seen as a logical next step after using descriptive analytics to spot trends (Cote, 2021). Diagnostic analytics can be applied in various ways in the retail sector:

a. *Analysis of Product Performance*: This type of analytics can help retailers understand why certain products perform better than others. By examining sales data, customer feedback, and market trends, retailers can pinpoint the factors contributing to a product's success or underperformance (Akter et al., 2022; Dahiya et al., 2022). Tools like machine learning algorithms and natural language processing can help analyze customer reviews and feedback for deeper insights.
b. *Customer Segmentation*: Retailers can use diagnostic analytics to understand different customer groups' distinct characteristics and behaviors. This understanding can help customize marketing and sales strategies to meet each segment's needs and preferences (Barua, 2018; Leão & da Silva, 2021; Lutfi

et al., 2023). Techniques such as clustering algorithms and decision trees are useful in identifying distinct customer segments.

c. *Evaluation of Marketing Campaigns*: Retailers can utilize diagnostic analytics to assess the effectiveness of their marketing efforts. By analyzing customer engagement, conversion rates, and sales data, retailers can identify the successful aspects of their campaigns and areas that need improvement (Khanom, 2023; Martin & Palmatier, 2020). Tools like regression analysis and A/B testing are commonly used in this evaluation.

   Moreover, personalized marketing campaigns can be developed based on segmentation. For instance, customers who frequently purchase children's clothing might receive promotions for back-to-school sales, while customers who often buy gourmet food items might receive recipes and cooking tips.

   Additionally, retailers can use data analytics to personalize each customer's shopping experience. An online retailer might display a customized homepage for each customer, featuring products the customer is likely to be interested in based on past purchases and browsing behavior.

4.3 Predictive Analytics

Predictive Analytics uses statistical models and forecasting methods to gain insight into future outcomes. This form of analytics helps retailers anticipate customer behavior, sales trends, and market changes. Predictive analytics can inform retail decisions, from inventory management to marketing strategies. Predictive analytics can be used in various ways in the retail sector:

a. *Forecasting Sales*: Predictive analytics can help retailers forecast future sales based on historical data and trends. This can inform inventory management decisions, helping prevent stock-outs and overstock situations (Akter et al., 2022; Dahiya et al., 2022).
b. *Predicting Customer Behavior*: Retailers can use predictive analytics to anticipate customer behavior. For instance, by analyzing past purchase history and browsing behavior, retailers can predict the products a customer will likely buy. This can inform personalized marketing strategies and recommendations (Akter et al., 2022; Devarapalli, 2023; Marr, 2021; Pressley, 2024).
c. *Anticipating Market Changes:* Predictive analytics can help retailers anticipate changes. By analyzing market trends and competitor activities, retailers can predict future changes in the market and adjust their strategies accordingly. Barua (2018) highlighted the application of predictive analytics in her study of European retail firms. Papanagnou et al. (2022) investigated the role of data-driven digital transformation in emergency scenarios by embracing a dynamic capability perspective. They examined the opinions of key stakeholders on the proactive employment of big data capabilities of UK grocery retail stores and the related

predictive analytics tools and methods. They found that data-driven digital transformation can assist businesses in predicting and swiftly reacting to disruptions, but big data capabilities were not crucial for enhancing supply chain resilience.

d. *Risk Analysis*: Predictive analytics can be used in retail risk analysis. For example, retailers can identify potential risks and take preventive measures by analyzing patterns in returns and fraudulent transactions (Araz et al., 2020).

Amazon, a prime example of a retailer that effectively uses data analytics, employs predictive analytics to recommend products to customers based on their browsing and purchasing history. It is worth noting that Amazon's recommendation engine drives 35% of its total sales (Marr, 2021). Target, a well-known retailer, also uses predictive analytics to anticipate customer behavior and market changes. For instance, they developed a pregnancy prediction model to send targeted ads to customers who were likely to be expecting (Julie, 2015). Although not a traditional retailer, Netflix, a streaming service, uses data analytics to personalize the customer experience. They use predictive analytics to recommend shows and movies based on a customer's viewing history (Devarapalli, 2023).

4.4 Prescriptive Analytics

Prescriptive analytics utilizes optimization and simulation algorithms to offer suggestions for potential outcomes. It goes beyond predicting future results by suggesting actions to capitalize on predictions and illustrating the implications of each decision option to the decision made. In the retail sector, prescriptive analytics can aid in optimizing pricing, discounts, and supply chain processes. There are numerous uses for prescriptive analytics within the retail sector:

a. *Optimizing Pricing and Discounts*: Digital technology has allowed for more advanced pricing strategies, such as dynamic pricing, which involves adjusting prices based on supply and demand in real-time. Retailers can utilize prescriptive analytics to optimize pricing and discounts by analyzing customer behavior, market conditions, and competitor pricing. Retailers can adapt their prices in real-time to attract customers and increase sales (Pressley, 2024).

   Digital technologies have also enabled the implementation of more sophisticated pricing strategies in retail networks. For example, retailers can use omnichannel models for customer service, allowing them to adjust pricing strategies based on customer behavior across different channels (Alexandrova & Kochieva, 2021). Moreover, digital technologies have transformed customer value in retail network services, enabling retailers to justify higher prices by enhancing customer value through digital technologies, thereby increasing profits (Chernova et al., 2020).

   Therefore, data analytics and digital technologies play essential roles in shaping pricing strategies in the retail sector. By leveraging these tools, retailers can establish competitive prices that maximize profits and enhance customer satisfaction and loyalty.

b. *Improving Supply Chain Processes*: By utilizing prescriptive analytics, retailers can optimize their supply chain processes by analyzing data on supplier performance, transportation costs, and delivery times. This helps identify bottlenecks and allows corrective action (Leão & da Silva, 2021; Pinto et al., 2023; Rossato & Castellani, 2020).
c. *Forecasting Product Demand*: Prescriptive analytics can forecast product demand based on historical sales and seasonal trends, enabling retailers to maintain optimal stock levels and reduce overstock costs (Akter et al., 2022).
d. *Enhancing Customer Experiences*: Retailers can enhance customer experiences by creating personalized shopping experiences using prescriptive analytics. For example, an online retailer might display a personalized homepage for each customer, featuring products the customer is likely to be interested in based on past purchases and browsing behavior. Nike's digital transformation strategy centers around personalized experiences. Through their Nike Training Club app, they offer tailored workout plans, while their Nike By You platform empowers customers to design their sneakers. By harnessing data and personalization, Nike enhances brand loyalty and engagement (Nike, 2024).
e. *Informing Strategic Decisions*: Prescriptive analytics can help inform strategic decisions in retail, such as inventory management and marketing strategies. For instance, if more than 50% of customers in a dataset reported being "very unsatisfied" with the customer service team, the algorithm could suggest additional training. Dahiya et al. (2022) emphasized the strategic importance of firm-specific knowledge in big data analytics and gaining a competitive advantage.

The different forms of data analysis, Descriptive, Diagnostic, Predictive, and Predictive- each play a distinct role in the retail sector. They can provide a comprehensive understanding of the data, enabling informed decision-making (Cote, 2021).

Data analytics is crucial for gaining a competitive advantage in the retail industry. By utilizing these four forms of data analytics, retailers can make data-driven decisions, foresee market trends, and improve customer experiences. For example, Descriptive Analytics helps retailers understand past performance; Diagnostic Analytics provides insights into the reasons for specific outcomes; Predictive Analytics forecasts future outcomes; and Prescriptive Analytics suggests actions for optimal results.

These analytics capabilities enable retailers to quickly adapt to changing market dynamics, better understand their customer base, and effectively tailor their strategies. This comprehensive approach to data analytics drives competitive advantage by informing strategic decisions, optimizing operations, and creating personalized customer experiences (Akter et al., 2022; Bresciani et al., 2021; Dahiya et al., 2022; Garg & Goyal, 2019; Hänninen et al., 2018; Khanom, 2023; Kumar, 2024; Leão & da Silva, 2021; Ramazanov et al., 2021; Rossato & Castellani, 2020; Shah, 2022).

4.5 Tesco Case Study

In this section, we will examine a case study of Tesco, a leading retailer, to gather insightful lessons and observations on how Tesco has successfully leveraged data analytics in its digital transformation journey.

A global retail leader, **Tesco** has been navigating the digital transformation wave by incorporating state-of-the-art technologies to maintain its competitive advantage (Research & Markets, 2024). This case study explores Tesco's strategic initiatives in the digital landscape, focusing on its commitment to digitization, innovation, and enhancing customer experiences.

Tesco's digital transformation journey is characterized by a progressive approach with technology at the core. The company uses advanced AI to optimize customer experiences, improve demand forecasting, analyze consumer behaviors, and prevent fraudulent activities. Tesco continues modernizing its stores with digital solutions that empower customers and streamline shopping processes (Research & Markets, 2024). A cornerstone of Tesco's strategy is the creation of a comprehensive data infrastructure, transitioning to more advanced and versatile systems like Hadoop. This structural transformation facilitates data access across various business operations, catalyzing informed decision-making and fostering a culture of data literacy within the organization.

In 2023, Tesco initiated a digital transformation program for its UK cafés, collaborating with the technology solutions provider Givex. As part of this initiative, digital kiosks from Givex were installed in the cafés to streamline ordering, automate kitchen operations, manage inventory, and control menus. The primary objective of this initiative was to enhance productivity and elevate the customer experience (Grocery Gazette, 2023).

Tesco's ongoing investment in digital and technological infrastructure underscores its unwavering dedication to pioneering the digital revolution in retail (Research & Markets, 2024). Despite challenges, Tesco's strategic strength and flexibility give it an advantage in the dynamic retail sector (Grocery Gazette, 2023). In response to the 2020 pandemic outbreak, Tesco acted swiftly by expanding its online order capacity to over 1.5 million per week, marking an increase of more than 100%. To accommodate this surge, Tesco initiated the operation of urban fulfillment centers and compact, automated warehouses within each store. The retailer has plans to establish additional fulfillment centers in the upcoming 12 months to enhance its online delivery services further. Since the onset of the pandemic, Tesco has seen a substantial growth of 77% in its online sales (Harvard Business Review, 2021).

A significant aspect of Tesco's digital transformation journey is its Clubcard loyalty program. This program collects customer data to offer personalized deals, enhancing customer retention (Yawson & Yamoah, 2022). The use of data analytics to refine this program and foster customer loyalty is a noteworthy aspect of Tesco's strategy.

In conclusion, Tesco's digital transformation journey is a compelling case study in the retail sector. The company's strategic use of advanced technologies and data analytics has enhanced customer experiences and provided a competitive advantage in the dynamic retail landscape. Despite the challenges posed by the digital revolution and the pandemic, Tesco's agility and adaptability have been key to its success. The company's commitment to innovation and investment in digital infrastructure and data literacy underscores the pivotal role of digital transformation in today's retail industry. Tesco's story is a testament to the power of data-driven decision-making and the potential of digital technologies to redefine customer experiences and drive business growth. It provides valuable insights and lessons for other retailers embarking on digital transformation journeys.

5 Challenges and Opportunities in Digital Transformation in Retail

The advancement of digital technologies continually impacts the retail industry, leading to significant and sometimes drastic shifts in the competitive landscape (Ferreira et al., 2020).

Despite the numerous benefits of digital transformation and data analytics for retailers, they also come with several challenges. These challenges include concerns about data privacy and security, the requirement for substantial investment in technology and skills, and the difficulty of integrating new technologies with existing systems (Dekimpe, 2020).

Abbu and Gopalakrishna (2022) studied the utilization of big data analytics by significant grocery businesses such as Walmart, Kroger, Albertsons, Target, and Amazon Whole Foods for digital transformation. The research highlights the crucial role of big data in various aspects of digital transformation. The grocery industry, known for intense competition and low margins, faces challenges and opportunities due to digital transformation. Retailers are investing in enhancing customer loyalty, ultimately impacting revenue and profit. Van Dyk and Van Belle (2020) highlighted significant obstacles to digital transformation adoption in the retail sector, including high implementation costs, legacy IT systems that are not easily adaptable to new technologies, and organizational resistance to change.

Reinartz et al. (2019) stress that consumers' purchase decisions were traditionally heavily influenced at the store shelf, giving brick-and-mortar retailers significant power to understand and shape consumer behaviors and preferences. However, the rise of e-commerce, mobile shopping, and, more recently, smart technologies have brought in new competitors, challenging this long-standing dominance. This shift highlights the dynamic nature of the retail industry and the necessity for businesses to adapt and innovate in response to these changes.

5.1 Challenges in Digital Transformation in Retail

Implementing data analytics in the retail sector can provide numerous benefits, from improved customer experiences to more efficient operations. However, it also presents several challenges that retailers must overcome to leverage the potential of data analytics fully. This section discusses some of the key challenges as follows:

1. *Data Quality Issues.* Ensuring data quality is a fundamental challenge in executing data analytics. Retailers often deal with vast amounts of data from various sources, which can sometimes be incomplete, inconsistent, or outdated. Poor data quality can lead to inaccurate analysis and misguided decisions (Aversa et al., 2021; Fan & Geerts, 2022; Lutfi et al., 2023).
2. *Organizational Resistance.* Another obstacle is the organization's resistance to change. Implementing data analytics often requires significant changes in business processes and work practices, and employees may resist these changes due to fear of the unknown or concern about job security (Bejta, 2021; Van Dyk & Van Belle, 2020; Vial, 2019).
3. *Lack of Skills and Expertise.* Implementing data analytics requires specific skills and expertise in data science and machine learning. Many retailers lack these skills in-house and may find it challenging to recruit and retain the necessary talent (Ramazanov et al., 2021; Vial, 2019).
4. *Technological Infrastructure.* Retailers require a strong technological infrastructure to support data analytics effectively. This includes hardware, software, and network capabilities. The lack of such infrastructure can hinder the implementation of data analytics initiatives (Kane, 2017; Van Dyk & Van Belle, 2020).
5. *Data Privacy and Security.* The rise of data analytics in retail brings ethical issues, especially around data privacy and security. Retailers have an ethical duty to protect customer data and use it respectfully. This involves transparency about data usage, providing customers with clear choices, and robust security measures. Ethically, retailers should ensure fair practices in using customer data for personalized marketing. Navigating these ethical issues is key to building customer trust and enhancing brand reputation (Martin & Palmatier, 2020; Martin et al., 2020).
6. *Cultural Change.* Implementing data analytics often requires a cultural change within the organization. Retailers need to foster a data-driven culture where employees understand the value of data, are comfortable using data to make decisions, and are committed to maintaining data quality and integrity (Kane, 2017; Vial, 2019).
7. *Customer Acceptance.* Customer acceptance is another challenge in implementing data analytics in retail. Retailers need to ensure that their use of data analytics is transparent and that customers' privacy is respected. This involves communicating with customers about their data use and obtaining consent (Martin & Palmatier, 2020; Martin et al., 2020).

8. *Scalability.* Ensuring that data analytics solutions can scale to handle increasing volumes of data and support growing business needs is a significant challenge in retail (Cakir et al., 2021).

In conclusion, the digital transformation in the retail sector, mainly through data analytics, is challenging. From data quality issues to the need for cultural change within the organization, retailers must navigate a complex landscape to leverage data analytics's benefits fully. However, by understanding and addressing these challenges, retailers can unlock the full potential of data analytics, driving improved decision-making, enhanced customer experiences, and, ultimately, business growth. As the retail sector continues to evolve, the ability to effectively harness data analytics will be a key differentiator for success.

5.2 Opportunities in Digital Transformation in Retail

Data analytics presents numerous opportunities for retailers despite the challenges they face:

1. *Enhanced Customer Insights.* Data analytics provides deep insight into customer behavior, preferences, and needs, allowing retailers to offer personalized products and services and enhance customer experiences (Chernova et al., 2020; Pressley, 2024; Reinartz et al., 2019; Yawson & Yamoah, 2022).
2. *Improved Operational Efficiency.* By analyzing data related to sales, inventory, supply chain, and other operational areas, retailers can identify inefficiencies, make improvements, and achieve cost savings and increased profitability (Coresight Research, 2019; Leão & da Silva, 2021; Pinto et al., 2023; Rossato & Castellani, 2020).
3. *Predictive Capabilities.* Data analytics enables retailers to forecast future trends and make proactive decisions. For example, predictive analytics can forecast demand for different products, enabling retailers to optimize their inventory and avoid stockouts or overstocks (Akter et al., 2022; Papanagnou et al., 2022).
4. *Strategic Decision-Making.* With big data, retailers can make strategic decisions that lead to a sustainable competitive advantage. This includes decisions related to market positioning, resource allocation, and strategic initiatives (Dahiya et al., 2022; Hagiu & Wright, 2020; Hänninen et al., 2018; Khanom, 2023; Kumar, 2024; Stephenson, 2018).
5. *Digital Transformation.* Data analytics is a crucial driving force behind digital transformation in the retail industry. By embracing digital technologies, retailers can innovate their business models, improve their operations, and enhance their customer experiences (Cakir et al., 2021; Malenkov et al., 2021; Papanagnou et al., 2022; Proskurnina et al., 2021; Rossato & Castellani, 2020).
6. *Supply Chain Management.* Data analytics can transform supply chain management in retail, optimize supply chains, improve inventory management, and

enhance logistics operations (Leão & da Silva, 2021; Pinto et al., 2023; Rossato & Castellani, 2020).

7. *Customer Value Creation.* Through data analytics, retailers can better understand and respond to customer needs, creating superior customer value, leading to increased customer loyalty and higher sales (Bresciani et al., 2021; Chernova et al., 2020; Hännine et al., 2018; Rossato & Castellani, 2020; Zaki, 2019).
8. *Reduction in Return Rates:* By analyzing patterns in return behavior, retailers can identify and address common issues, such as inaccurate product descriptions or sizing problems. This can significantly reduce return rates and associated costs, ultimately impacting overall profitability (Akter et al., 2022; Ferreira et al., 2020).
9. *Cost Reduction*: Data-driven decisions can streamline operations and reduce costs. For instance, predictive maintenance can prevent equipment failures and reduce downtime, while inventory optimization can minimize storage costs (Araz et al., 2020; Pinto et al., 2023).

In conclusion, despite the challenges, implementing data analytics in retail offers significant opportunities for retailers willing to navigate these challenges and leverage data for competitive advantage.

6 Future Trends in Digital Transformation in Retail

The future of data analytics in retail is promising, with revolutionary changes expected to transform the industry in many ways. In this section, we will discuss some of the emerging trends and technologies that will likely shape the future of retail.

Artificial Intelligence and Machine Learning: Akter et al. (2022) explore the transformative impact digital innovations, including artificial intelligence and machine learning, can have on the business sector. These technologies are expected to play a significant role in retail data analytics, providing new ways to gather, analyze, and secure data. AI-driven tools can enhance predictive analytics, automate customer service through chatbots, and optimize supply chain management. This could lead to more efficient operations, improved customer service, and enhanced security.

Blockchain Technology: As highlighted by Akter et al. (2022), Blockchain technology can revolutionize data security and transparency in retail. It allows for secure, transparent transactions and can help track the provenance of products, ensuring authenticity and reducing fraud. This technology can enhance supply chain transparency, build customer trust, and streamline operations.

*Omnichannel Retailing***:** Cakir et al. (2021) highlight the role of omnichannel retailing in a medium-sized retailer's digital transformation. With the growth of e-commerce, retailers will need to leverage data analytics to create seamless shopping experiences across multiple channels, including online, in-store, and mobile. This approach can increase customer satisfaction and loyalty by providing a consistent and personalized shopping experience across all touchpoints.

Digital Transformation Strategy: Bresciani et al. (2021) emphasize the importance of building a comprehensive digital transformation strategy. As retailers continue adopting data analytics, they must strategically plan their digital transformation journey. This includes considering organizational culture, technology infrastructure, and employee skills. Retailers who successfully navigate their digital transformation will likely gain a competitive advantage by being more agile, customer-centric, and innovative.

Internet of Things: IoT devices can collect real-time data from various sources, such as smart shelves and sensors, to provide valuable insights into inventory levels, customer behavior, and store operations. This real-time data can help retailers optimize inventory management, enhance the in-store experience, and reduce operational costs.

Augmented Reality (AR) and Virtual Reality (VR): AR and VR technologies can significantly enhance the customer shopping experience. AR can create interactive product displays and virtual try-ons, while VR can offer immersive shopping experiences. These technologies can help bridge the gap between online and in-store shopping, making the shopping experience more engaging and personalized.

Data Privacy and Ethical Considerations: As data analytics grows, ensuring data privacy and ethical use of customer data will become increasingly important. Retailers must comply with regulations such as the General Data Protection Regulation (GDPR) and adopt the best data security and privacy practices. Transparent data practices will help build customer trust and loyalty.

These trends underscore the industry's resolve to adapt, innovate, and maintain competitiveness in a world after COVID-19. As these technologies evolve, they will undoubtedly bring about significant changes in retail operations and consumer experiences. Retailers must stay updated with these trends and be prepared to leverage these opportunities fully.

7 Conclusions

The digital transformation in the retail sector, primarily through integrating and applying data analytics, offers retailers a competitive advantage by enabling informed decision-making and enhancing operational efficiency. This chapter has explored how data analytics can be leveraged in retail, from understanding sales trends and customer behavior to optimizing inventory management and pricing strategies.

However, implementing data analytics in retail has its challenges. These challenges encompass data privacy and security issues, the need for significant investment in technology and expertise, and the difficulty of integrating new technologies with existing systems. It is crucial for retailers to effectively address these challenges to unlock the full potential of data analytics.

Despite these challenges, retailers have numerous potential opportunities. These include the ability to reach a global audience, the creation of highly targeted marketing campaigns, and the capacity to gather and analyze vast amounts of customer data. The future of data analytics in retail is promising, with emerging trends and technologies poised to revolutionize the sector further.

This chapter has highlighted several real-world examples illustrating how data analytics can provide a competitive advantage in the retail sector. Companies like Tesco, Amazon, Walmart, Target, and Netflix have successfully leveraged data analytics techniques to gain a competitive advantage, demonstrating the transformative potential of data analytics in retail.

In conclusion, data analytics plays a crucial role in the retail sector. By leveraging data analytics, retailers can make data-driven decisions, anticipate market trends, enhance customer experiences, and gain a competitive advantage. As these technologies evolve, they will significantly change retail operations and consumer experiences.

However, retailers must stay updated with these trends and be prepared to leverage these opportunities fully. While these trends offer many opportunities, it is crucial to address the challenges effectively to reap the full benefits.

Therefore, retailers must embrace digital transformation and data analytics not as an option but as a strategic necessity to stay competitive and relevant in the digital age. The future of retail will be shaped by those who can effectively navigate these challenges and harness the power of data analytics to drive decision-making, personalize customer experiences, and optimize operations.

This chapter contributes to the scholarly discourse on digital transformation and data analytics in the retail sector. It provides valuable insights for academics, researchers, and industry professionals interested in this field.

References

Abbu, H., & Gopalakrishna, P. (2022). Digital transformation powered by big data analytics: The case of retail grocery business. In *Proceedings of the 55th Hawaii International Conference on System Sciences*. https://doi.org/10.24251/hicss.2022.231

Akter, S., Michael, K., Uddin, M. R., McCarthy, G., & Rahman, M. (2022). Transforming business using digital innovations: The application of AI, blockchain, cloud and data analytics. *Annals of Operations Research, 308*, 7–39. https://doi.org/10.1007/s10479-020-03620-w

Alexandrova, E., & Kochieva, A. (2021). Modern aspects of digital technologies development in retail networks. In *Comprehensible Science: ICCS 2020* (pp. 111–120). Springer International Publishing. https://doi.org/10.1007/978-3-030-66093-2_11

Araz, O. M., Choi, T. M., Olson, D. L., & Salman, F. S. (2020). Role of analytics for operational risk management in the era of big data. *Decision Sciences, 51*(6), 1320–1346. https://doi.org/10.1111/deci.12451

Aversa, J., Hernandez, T., & Doherty, S. (2021). Incorporating big data within retail organizations: A case study approach. *Journal of Retailing and Consumer Services, 60*, Article 102447. https://doi.org/10.1016/j.jretconser.2021.102447

Barua, B. (2018). *Gaining competitive advantage* in *retail firms with the help of digital platform for segmentation process–A case study on European retail firms*. Doctoral dissertation, Dublin, National College of Ireland. https://norma.ncirl.ie/3388/1/bhaswatibarua.pdf

Bejta, M. (2021). *Surviving digital disruption: identifying the key digital transformation barriers of multi-channel retailers.* Master's thesis, University of Twente. Retrieved June 18, 2024, from https://essay.utwente.nl/85627/7/Bejta_MA_BMS.pdf

Bresciani, S., Ferraris, A., Romano, M., & Santoro, G. (2021). Building a digital transformation strategy. In *Digital transformation management for agile organizations: A compass to sail the digital world* (pp. 5–27). Emerald Publishing Limited. https://doi.org/10.1108/978-1-80043-171-320211002

Cakir, G., Iftikhar, R., Bielozorov, A., Pourzolfaghar, Z., & Helfert, M. (2021). Omnichannel retailing: Digital transformation of a medium-sized retailer. *Journal of Information Technology Teaching Cases, 11*(2), 122–126. https://doi.org/10.1177/2043886920959803

Chernova, D. V., Sharafutdinova, N. S., Nurtdinov, I. I., Valeeva, Y. S., & Kuzmina, L. I. (2020). The transformation of the customer value of retail network services under digitalization. In *Digital age: Chances, challenges and future 7* (pp. 252–260). Springer International Publishing. https://doi.org/10.1007/978-3-030-27015-5_31

Chorna, M., Bezghinova, L., Dorokhov, O., Zhuvahina, I., & Volosov, A. (2021). Efficiency of retail enterprises in context of achievement of competitive advantages: Ukrainian realities. *TEM Journal, 10*(3), 1072–1081. https://doi.org/10.18421/tem103-10

Coresight Research (2019). *How data analytics is driving digital transformation in retail.* Retrieved June 15, 2024, from https://coresight.com/research/how-data-analytics-is-driving-digital-transformation-in-retail/

Cote, C. (2021). *4 types of data analytics to improve decision-making.* Retrieved May 5, 2024, from https://online.hbs.edu/blog/post/types-of-data-analysis

Dahiya, R., Le, S., Ring, J. K., & Watson, K. (2022). Big data analytics and competitive advantage: The strategic role of firm-specific knowledge. *Journal of Strategy and Management, 15*(2), 175–193. https://doi.org/10.1108/jsma-08-2020-0203

Dekimpe, M. G. (2020). Retailing and retailing research in the age of big data analytics. *International Journal of Research in Marketing, 37*(1), 3–14. https://doi.org/10.1016/j.ijresmar.2019.09.001

Devarapalli S. (2023). *Breaking bad with customer analytics: A Netflix perspective.* Retrieved May 12, 2024, from https://datazip.io/blog/breaking-bad-with-customer-analytics-a-netflix-perspective/

Fan, W., & Geerts, F. (2022). Foundations of data quality management. *Springer Nature.* https://doi.org/10.1007/978-3-031-01892-3

Ferreira, M. J., Moreira, F., Pereira, C. S., & Durão, N. (2020). The digital transformation at organizations–the case of retail sector. In *Trends and innovations in information systems and technologies: Volume 1 8* (pp. 560–567). Springer International Publishing. https://doi.org/10.1007/978-3-030-45688-7_564

Garg, A., & Goyal, D. P. (2019). Sustained business competitive advantage with data analytics. *International Journal of Business and Data Analytics, 1*(1), 4–15. https://doi.org/10.1504/ijbda.2019.10020198

Grocery Gazette (2023). *Tesco launches digital transformation programme across cafés.* Retrieved May 12, 2024, from https://www.grocerygazette.co.uk/2023/07/19/tesco-digital-transform-cafes/

Hagiu A. & Wright J. (2020). *When data creates competitive advantage.* Retrieved May 5, 2024, from https://hbr.org/2020/01/when-data-creates-competitive-advantage

Hänninen, M., Smedlund, A., & Mitronen, L. (2018). Digitalization in retailing: Multi-sided platforms as drivers of industry transformation. *Baltic Journal of Management, 13*(2), 152–168. https://doi.org/10.1108/bjm-04-2017-0109

Harvard Business Review (2021). *How disruption accelerated digital supply chain transformation.* Retrieved May 11, 2024, from https://hbr.org/sponsored/2021/08/how-disruption-accelerated-digital-supply-chain-transformation

He, T., Liu, W., Shao, X., & Tian, R. G. (2023). Exploring the digital innovation process and outcome in retail platform ecosystems: Disruptive transformation or incremental change. *Electronic Commerce Research,* 1–30. https://doi.org/10.1007/s10660-023-09699-0

Insights Desk (2023). *The power of retail data analytics: Understanding its role and the different types*. Retrieved June 10, 2024, from https://www.demandtalk.com/insights/data/big-data/the-power-of-retail-data-analytics-understanding-its-role-and-the-different-types/

Julie (2015). *Target using predictive analytics to increase value capture*. Retrieved May 13, 2024, from https://d3.harvard.edu/platform-digit/submission/target-using-predictive-analytics-to-increase-value-capture/

Kane, G. C. (2017). Digital maturity, not digital transformation. *MIT Sloan Management Review, 1*(1), 1–15. https://sloanreview.mit.edu/article/digital-maturity-not-digital-transformation/

Khan, S. N., Sultan, M. F., & Asif, M. (2022). Association of big data with sustainable competitive advantage in online retail segment: A serial mediation model for relating big data with strategic management tools. *Reviews of Management Sciences, 4*(1), 168–180. https://doi.org/10.53909/rms.04.01.0138

Khanom, M. T. (2023). Business strategies in the age of digital transformation. *Journal of Business, 8*(01), 28–35. https://doi.org/10.18533/job.v8i01.296

Kothari, U., & Seetharaman, A. (2020). Impact of digital transformation on retail banking industry in the UAE. In *Re-imagining Diffusion and Adoption of Information Technology and Systems: A Continuing Conversation: IFIP WG 8.6 International Conference on Transfer and Diffusion of IT, TDIT 2020, Tiruchirappalli, India, December 18–19, 2020, Proceedings, Part I* (pp. 425–438). Springer International Publishing. https://doi.org/10.1007/978-3-030-64849-7_38

Kumar. S. (2024), Retail operations strategy for improved customer experience: a better response to crises such as COVID-19 pandemic, *Journal of Global Operations and Strategic Sourcing*, Vol. ahead-of-print No. ahead-of-print. https://doi.org/10.1108/JGOSS-05-2023-0044

Leão, P., & da Silva, M. M. (2021). Impacts of digital transformation on firms' competitive advantages: A systematic literature review. *Strategic Change, 30*(5), 421–441. Portico. https://doi.org/10.1002/jsc.2459

Levallet, N., & Chan, Y. E. (2018). Role of digital capabilities in unleashing the power of managerial improvisation. *MIS Quarterly Executive, 17*(1), 1–21.

Lutfi, A., Alrawad, M., Alsyouf, A., Almaiah, M. A., Al-Khasawneh, A., Al-Khasawneh, A. L.,... & Ibrahim, N. (2023). Drivers and impact of big data analytic adoption in the retail industry: A quantitative investigation applying structural equation modeling. *Journal of Retailing and Consumer Services, 70*, Article 103129. https://doi.org/10.1016/j.jretconser.2022.103129

Malenkov, Y., Kapustina, I., Kudryavtseva, G., Shishkin, V. V., & Shishkin, V. I. (2021). Digitalization and strategic transformation of retail chain stores: Trends, impacts, prospects. *Journal of Open Innovation: Technology, Market, and Complexity, 7*(2), 108. https://doi.org/10.3390/joitmc7020108

Marr, B. (2021). *How amazon uses artificial intelligence: The Flywheel approach*. Retrieved May 15, 2024, from https://bernardmarr.com/how-amazon-uses-artificial-intelligence-the-flywheel-approach/

Martin, K. D., & Palmatier, R. W. (2020). Data privacy in retail: Navigating tensions and directing future research. *Journal of Retailing, 96*(4), 449–457. https://doi.org/10.1016/j.jretai.2020.10.002

Martin, K. D., Kim, J. J., Palmatier, R. W., Steinhoff, L., Stewart, D. W., Walker, B. A., … & Weaven, S. K. (2020). Data privacy in retail. *Journal of Retailing, 96*(4), 474–489. https://doi.org/10.1016/j.jretai.2020.08.003

McKinsey & Company (2020). *Future of retail operations: Winning in a digital era*. Retrieved June 16, 2024, from https://www.mckinsey.com/~/media/McKinsey/Industries/Retail/Our%20Insights/Future%20of%20retail%20operations%20Winning%20in%20a%20digital%20era/McK_Retail-Ops-2020_FullIssue-RGB-hyperlinks-011620.pdf

Nike (2024). Retrieved June 20, 2024, from https://www.nike.com/nike-by-you

Pantano, E. (Ed.). (2020). *Retail futures: The good, the bad and the ugly of the digital transformation*. Emerald Publishing Limited. https://doi.org/10.1108/9781838676636

Papanagnou, C., Seiler, A., Spanaki, K., Papadopoulos, T., & Bourlakis, M. (2022). Data-driven digital transformation for emergency situations: The case of the UK retail sector. *International Journal of Production Economics, 250*, Article 108628. https://doi.org/10.1016/j.ijpe.2022.108628

Pinto, M. R., Salume, P. K., Barbosa, M. W., & de Sousa, P. R. (2023). The path to digital maturity: A cluster analysis of the retail industry in an emerging economy. *Technology in Society, 72*, Article 102191. https://doi.org/10.1016/j.techsoc.2022.102191

Pressley, J.P. (2024). To predict consumer behavior, retailers turn to data analytics. *BizTech Magazine.* Retrieved June 19, 2024, from https://biztechmagazine.com/article/2024/01/predict-consumer-behavior-retailers-turn-data-analytics

Proskurnina, N. V., Shtal, T. V., Slavuta, O. I., Serogina, D. O., & Bohuslavskyi, V. V. (2021). Omnichannel Strategy of digital transformation of retail trade enterprise: From concept to implementation. *Studies of Applied Economics, 39*(6), 1–10. https://doi.org/10.25115/eea.v39i6.5238

Quinones, M., Gomez-Suarez, M., Cruz-Roche, I., & Díaz-Martín, A. M. (2023). Technology: A strategic imperative for successful retailers. *International Journal of Retail & Distribution Management, 51*(4), 546–566. https://doi.org/10.1108/ijrdm-03-2022-0088

Ramazanov, I. A., Panasenko, S. V., Cheglov, V. P., Krasil'nikova, E. A. E., & Nikishin, A. F. (2021). Retail transformation under the influence of digitalisation and technology development in the context of globalisation. *Journal of Open Innovation: Technology, Market, and Complexity, 7*(1), 49. https://doi.org/10.3390/joitmc7010049

Reinartz, W., Wiegand, N., & Imschloss, M. (2019). The impact of digital transformation on the retailing value chain. *International Journal of Research in Marketing, 36*(3), 350–366. https://doi.org/10.1016/j.ijresmar.2018.12.002

Research and Markets (2024). Tesco retail digital transformation: In-depth insights on Tesco Plc's tech-driven strategies and innovation endeavors. Retrieved May 10, 2024, from https://finance.yahoo.com/news/tesco-retail-digital-transformation-depth-143800713.html

Rossato, C., & Castellani, P. (2020). The contribution of digitalisation to business longevity from a competitiveness perspective. *The TQM Journal, 32*(4), 617–645. https://doi.org/10.1108/TQM-02-2020-0032

Schneider, S., & Kokshagina, O. (2021). Digital transformation: What we have learned (thus far) and what is next. *Creativity and Innovation Management, 30*(2), 384–411. https://doi.org/10.1111/caim.12414

Shah, T. R. (2022). Can big data analytics help organisations achieve sustainable competitive advantage? A developmental enquiry. *Technology in Society, 68*, Article 101801. https://doi.org/10.1016/j.techsoc.2021.101801

Stephenson, D. (2018). *Big data Demystified: How to use big data, data science and AI to make better business decisions and gain competitive advantage.* Pearson UK.

Torres, R. (2022). How Walmart enhances its inventory, supply chain through AI. Retrieved May 15, 2024, from https://www.ciodive.com/news/walmart-AI-ML-retail/638582/

van Dyk, R., & Van Belle, J. P. (2020). Drivers and challenges for digital transformation in the South African retail industry. In *Information Technology for Management: Current Research and Future Directions: 17th Conference, AITM 2019, and 14th Conference, ISM 2019, Held as Part of FedCSIS, Leipzig, Germany, September 1–4, 2019, Extended and Revised Selected Papers 17* (pp. 42–62). Springer International Publishing. https://doi.org/10.1007/978-3-030-43353-6_3

Vial, G. (2019). Understanding digital transformation: A review and a research agenda. *Journal of Strategic Information Systems, 28*(2), 118–144. https://doi.org/10.1016/j.jsis.2019.01.003

Yawson, D. E., & Yamoah, F. A. (2022). The TESCO club card loyalty programme: The gold standard. In *Contemporary retail marketing in emerging economies: The case of Ghana's supermarket chains* (pp. 77–87). Springer International Publishing. https://doi.org/10.1007/978-3-031-11661-2_3

Zaki, M. (2019). Digital transformation: Harnessing digital technologies for the next generation of services. *Journal of Services Marketing, 33*(4), 429–435. https://doi.org/10.1108/jsm-01-2019-0034

Towards Customer Experience Enhancement Using Customer Analytics: A Case Study of a Leading Indian Private Bank

Rupali Anand Shirsat, Shilpa Desai, and Dinesh Sharma

Abstract Data analytics is revolutionizing financial services, transforming how institutions understand, engage with, and serve their customers. This chapter examines the profound impact of data analytics through a case study of a leading Indian private sector bank, illustrating its practical applications and strategic benefits. Despite its rapid industry adoption, academic research on data analytics remains limited, with best practices still underdeveloped. Extant literature highlights the transformative power of data analytics in service innovation, particularly in advanced economies. However, there is a notable gap in studies focused on developing countries like India. This study's primary contribution is applying the theories of customer experience in a practical industry context and demonstrating how a progressive Indian bank strategically integrates data analytics to enhance customer experience. By exploring the bank's real-world applications and mechanisms, the study offers valuable insights for other banking institutions struggling to harness digital innovations, manage data explosions, and sustainably deliver value to customers.

Keywords Customer Analytics · Predictive Analytics · Customer Experience · Financial Services

1 Introduction

The shift towards digitalization (Zaki, 2019) has captured worldwide focus across different sectors due to the widespread adoption and growth of recent digital innovations brought by new contemporary technologies such as Internet of Things (IoTs), artificial intelligence (AI), data analytics, and machine learning (Cho et al., 2023).

The shift towards digitalization (Zaki, 2019) has captured worldwide focus across different sectors due to the widespread adoption of contemporary technologies such as Internet of Things (IoTs), artificial intelligence (AI), data analytics, and machine

R. A. Shirsat (✉) · S. Desai · D. Sharma
Shailesh J. Mehta School of Management, Indian Institute of Technology Bombay, Mumbai, India
e-mail: rupali.anand@iitb.ac.in

B. Basu et al. (eds.), *Data Analytics for Decision Making towards Business Excellence*, Uncertainty and Operations Research, https://doi.org/10.1007/978-981-95-2429-7_3

learning (Cho et al., 2023). Globally, the evolution of data analytics in financial services began with the adoption of CRM systems in the 1990s, providing banks with tools to understand customer behaviour. Fintech innovations accelerated after the 2008 financial crisis, emphasizing personalized services and risk management through advanced analytics.

In India, key milestones have defined this evolution:

- **Jan Dhan Yojana, 2014** (Pradhan Mantri Jan-Dhan Yojana (PMJDY)| National Portal of India, n.d.)**:** Financial inclusion initiatives that expanded the banking base to millions of previously unbanked citizens.
- **Unified Payments Interface (UPI), 2016** (Mathew, 2024): A game-changer in digital payments, enabling seamless peer-to-peer and merchant transactions.
- **Demonetization, 2016** (Kumari, 2019)**:** A catalyst for digital transactions and data-driven banking services.
- **Digital India Initiative** (Singh, 2022)**:** Accelerated the adoption of digital platforms across industries, including banking.

These milestones have collectively driven the integration of data analytics into Indian banking, enabling institutions to personalize offerings, enhance customer experiences, and improve operational efficiency. However, academic research in this area remains limited, particularly in the Indian context, which presents unique challenges such as legacy systems, regulatory constraints, and resource limitations. This study aims to fill this gap by examining how a progressive Indian private-sector bank has strategically adopted data analytics to enhance customer experience.

The world is experiencing a technological upheaval. Innovations like data analytics have transformed numerous sectors by facilitating better-informed choices, enhancing processes, and fostering creativity (Soltani Delgosha et al., 2021). It is used in different important areas, ranging from healthcare, retailing, and supply chain, and it has a significant role in the financial services industry, notably banking (Peltier et al., 2020; Frizzo-Barker et al., 2016).

Per the World Retail Banking Report by Capgemini (2024), the banking sector was experiencing a transformative shift from 2014 to 2023, as the rise of the global fintech revolution and the COVID-19 pandemic in 2020 accelerated the ongoing digital transformation. This shift pushed the banking institutions to adopt and integrate digital technologies such as cloud and data analytics.

The report by Deloitte (2021) named *How can banks thrive in a digital society?* Informs that financial institutions struggle with old-fashioned systems and strict regulations, whereas agile and forward-thinking fintech startups focus on improving the customer experience. To overcome the intense competition from new-age fintech startups, it becomes evident how banks adopt technologies and methods to remain competitive; upgraded core systems that offer adaptability and instant data analysis will lay a solid groundwork for ongoing innovation and a competitive edge. Enhancing the customer experience (CX) is the key to success (Chauhan et al., 2022; Wipro, 2021). Data analytics has emerged as a transformative force for delivering elevated and meaningful customer experiences in the financial services sector (Hajiheydari et al., 2021), reshaping how financial institutions understand, engage with, and serve their customers. The financial services landscape is swiftly evolving, driven by technological advancements and shifting consumer behaviours. In this dynamic environment, data analytics has become indispensable for financial institutions striving to remain ahead of the competition and drive business growth (Frizzo-Barker et al., 2016; Saggi & Jain, 2018; Hajiheydari et al., 2021).

Despite its rapid growth in recent years, research on data analytics is still in its infancy (Soltani Delgosha et al., 2021) and has not received as much attention as it deserves. Therefore, efforts to explore its strategic uses, key factors, and significant obstacles are crucial steps forward. Additionally, it is essential to prioritize these findings to guide practitioners and researchers in focusing on the most impactful aspects. While there has been extensive evaluation of the value of information technology in recent years, there is a demand for progress in data analytics value research. This challenge exists because the current knowledge of the best practices for deriving value from data analytics technologies is still quite limited (Côrte-Real et al., 2019).

In today's digital environment (Gurbaxani & Dunkle, 2019), the limits of creativity have significantly expanded. Historically, creativity was mainly linked to developing new products or services and, to a lesser degree, improving processes (Sorescu & Schreier, 2021). As a result of the growing standardization of goods and the escalating desire of consumers for personalized experiences and engagements are leading top executives to redirect their attention from developing new products to enhancing services (Lehrer et al., 2018; Vargo et al., 2020; Soto Setzke et al., 2023).

The extant literature suggests the huge revolutionary potentials brought by data analytics in service innovations, especially in the banking sector in developed and technologically advanced countries as compared to developing nations like India, the literature focussing on developing countries is very limited (Al-Dmour et al., 2023; Gambetta et al., 2016; Nguyen et al., 2024; Singh & Sharma, 2022).

The banking sector in India is experiencing a significant shift due to the fintech revolution in the past few years brought by digital technologies, and the various government schemes leading to financial inclusion have paved the path for massive growth of the banking sector in the country. As financial institutions keep using data and sophisticated analytical tools, they can provide personalized offerings to customers, advance their processes, and handle risks more efficiently. This evolution presents a critical opportunity for academicians and researchers to provide valuable insights backed by strong theoretical and conceptual underpinnings, as minimal studies in the literature have focused on the potential of data analytics in the Indian banking sector.

The main contribution of this study lies in describing the strategic integration of data analytics for delivering enhanced customer experience by a leading new-age Indian private sector bank. This study uses a case study approach (Yin, 2012; Stake, 1995; Yin, 2002) and considers India's leading private sector bank, which has innovatively and effectively transformed its customer experience by leveraging data analytics. Using Yin's approach to case study writing helped us to link theory and practice by presenting the width and breadth of case study research and its significance at a practical level.

Understanding the real-world application and the underlying mechanism used by the leading bank considered in this study provides insights for other banking institutions that are struggling to integrate these digital innovations (Davenport & Westerman, 2018) and are not able to cope with massive data explosion and missing providing the ultimate value to customers sustainably.

2 Overview—Theoretical Background

This section enunciates a brief conceptual overview of customer experience and CX in services, background of data analytics, and mentions the major theoretical underpinnings for understanding the underlying importance of customer experience in financial services, especially banking.

In recent years, significant advancements in data analytics for customer experience enhancement have occurred. Studies by Nguyen et al. (2024) and Al-Dmour et al. (2023) have demonstrated how emerging markets are beginning to adopt analytics for personalized banking. However, their focus is primarily on infrastructure rather than customer-centric innovations. Similarly, work by Tueanrat et al. (2021) highlights the role of customer journey mapping in banking analytics but is limited to developed markets. This study bridges this gap by focusing on how a progressive Indian private-sector bank integrates data analytics to optimize customer experiences.

2.1 Customer Experience Construct

The development of the idea behind the customer experience over time illuminates its conceptual beginnings (Lemon & Verhoef, 2016). The concept of customer experience has only recently become prominent in academic marketing theory and its application, with a significant increase in attention over the past three decades (Maklan & Klaus, 2011; Jain et al., 2017).

Customer experience is predominantly regarded as the marketing concept (Lemon & Verhoef, 2016; Jain et al., 2017; Becker & Jaakkola, 2020) and has received significant focus in academic research and real-world application for nearly fifty years. It is multidimensional (Lemon & Verhoef, 2016; Schmitt, 1999; Siebert et al., 2020) and multilevel (Patrı´Cio et al., 2011; Kranzbühler et al., 2018) dynamic phenomenon (McColl-Kennedy et al., 2015) that encompasses all the interactions a customer has with the company's offerings throughout the customer journey. The interactions can be through various touchpoints it could be in the form of advertising, features of a product/service, packaging, ease of use, or customer service (Kranzbühler et al., 2018; Lemon & Verhoef, 2016; Meyer & Schwager, 2007) spread across the customer journey stages (pre-purchase, purchase, or post-purchase) (Verhoef et al., 2009). This concept has steadily gained importance, evolving into a crucial factor for companies aiming to provide seamless and personalized customer experiences, gaining an advantage over rivals and securing success in the expanding arenas of digital and physical commerce (Bolton et al., 2018; De Keyser et al., 2015; Maklan & Klaus, 2011; Silva et al., 2021; Wetzels & Wetzels, 2023).

The foundational study by Holbrook and Hirschman (1982) highlighted the significance of the experiential elements of consumption, which play a crucial role in fulfilling consumer needs and desires. Meanwhile, Pine and Gilmore (1998) emphasized the necessity and value of creating unique experiences that differ from products or services. Schmitt (1999) provided a comprehensive perspective, categorizing experiences into five dimensions: sensory (perception), affective (emotional), cognitive (thought processes), physical (actions), and social identity (relationships). The research by Verhoef et al. (2009) marked a significant shift in customer experience studies. By building upon Holbrook and Hirschman's (1982) work, they aimed to elucidate the concept of customer experience within the retail sector, highlighting its comprehensive nature that includes cognitive, affective, emotional, social, and physical reactions to the store environment. Brakus et al. (2009) introduced the idea of brand experience as personal, internal reactions triggered by stimuli related to a brand. They outlined the four aspects of brand experience as sensory, affective, intellectual, and behavioural. Lemon and Verhoef (2016) further developed the concept of the multidimensional customer experience, focusing on consumers' cognitive, emotional, behavioural, sensory, and social responses to a company's offerings throughout the customer journey.

2.2 *Customer Experience in Services*

A customer's experience with services is a continuous, comprehensive journey that demands a thorough grasp of what customers need, seamless integration across various interactions, and a significant focus on personalized, empathetic service (Groth et al., 2019; Holmlund et al., 2020). Service providers who master these aspects are more inclined to build deep customer loyalty, positive recommendations, and sustained achievements (Frow & Payne, 2007). By consistently adjusting and enhancing their service provision in response to customer input and technological progress, they can craft an attractive and lasting customer experience that distinguishes them in a competitive landscape (Holmlund et al., 2020).

Customer experience (CX) in services is a multifaceted concept encompassing a customer's entire journey when interacting with a service provider (McColl-Kennedy et al., 2015). It is not confined to a single touchpoint but spans the entire customer lifecycle, from initial awareness to post-service interactions (Voorhees et al., 2017). In essence, CX in services is the sum of all customer perceptions, emotions, and reactions in response to the cumulative encounters with a service brand. It begins with the first impression, often shaped by marketing and initial inquiries, setting the tone for expectations (Vargo & Lusch, 2004). A positive customer experience is achieved through a seamless, personalized, and efficient service that meets or exceeds customer expectations, while a negative experience can stem from inefficiencies, lack of personalization, or unmet expectations (Bueno et al., 2019; Chahal & Dutta, 2015). Offering outstanding customer experiences has turned into a primary goal in today's contemporary service sector, ranging from retail (Grewal et al., 2009), hospitality and tourism (Kandampully et al., 2018), hotels (Alnawas et al., 2019; Rahimian et al., 2021), healthcare (Kumar, 2022) to banking (Chahal & Dutta, 2015; Klink et al., 2020; Maklan et al., 2017) experience of receiving services influences customer satisfaction, loyalty, and their willingness to recommend the service to others through spreading positive or negative Word of mouth about depending on their service experiences ("Phil" Klaus & Chahal & Dutta, 2015; Maklan, 2012).

It is evident from the above information that in today's digital world, customers expect to have access to a service experience (Gustafsson et al., 2024; "Phil" Klaus & Maklan, 2012; Voorhees et al., 2017) that is seamless, personalized, and delightful (Parasuraman et al., 2020). The shift in businesses and human life due to digital transformation (Zaki, 2019) as a result of the availability of disruptive digital technologies such as data analytics (Lehrer et al., 2018; Marinakos & Daskalaki, 2017; Moeyersoms & Martens, 2015) enables businesses to identify the pain points of customers throughout their customer journeys and customize the company's offerings as per the customers' expectations and have an edge in the highly volatile competitive battleground (Groth et al., 2019; Gustafsson et al., 2024; Zaki, 2019).

The transformative potential of data analytics has been explored in several contexts, particularly in developed economies. Studies by Frizzo-Barker et al. (2016) and Saggi and Jain (2018) demonstrate how predictive analytics can enhance customer segmentation, improve retention, and enable personalized marketing in banking. In advanced economies, real-time analytics has become a standard tool for improving customer service and operational efficiency (Hajiheydari et al., 2021). However, research in emerging markets is limited. While Al-Dmour et al. (2023) and Nguyen et al. (2024) provide insights into data analytics adoption in developing regions, they focus more on infrastructure and less on customer experience. This study addresses this gap by focusing on the strategic integration of data analytics to improve customer experiences in the Indian banking sector.

2.3 Theoretical Frameworks

The data analytics application for enhancing customer experience within banking can be comprehended through various significant theoretical models. The following significant theoretical frameworks are used for developing the underlying case study:

2.3.1 Customer Journey Mapping

Understanding the customer's journey has become a crucial idea for grasping intricate customer actions and gaining an understanding of their experiences across multiple touchpoints spread over customer journeys (Lemon & Verhoef, 2016; Micheaux &

Bosio, 2019; Tueanrat et al., 2021). Integration of data analytics in customers' banking journey will aid the banking institute in identifying the most critical touch-points and designing and offering the most effective services, which may benefit improved customer satisfaction, loyalty, and recommendations (Kuehnl et al., 2019; Siebert et al., 2020).

2.3.2 Customer Relationship Management (CRM)

The CRM model offers an essential structure for understanding the procedures banks can use to develop and preserve lasting associations with customers. At the heart of CRM lies the idea of focusing on customers (Ascarza et al., 2017), highlighting the significance of recognizing and meeting their needs and desires (Hillebrand et al., 2011). Data analytics plays a vital role in CRM by giving banks practical knowledge about how customers behave and allows for accurate grouping of customers by their behaviour, demographic characteristics, and psychological traits., serving for personalized engagements and focused promotional activities (Becker et al., 2009; Chang et al., 2010).

3 Case Study

3.1 Background

The merger of two entities, a corporate and infrastructure financing bank and a retail finance leader, created a unique opportunity to blend their strengths and capabilities. The corporate and infrastructure financing bank brought substantial expertise in large-scale project financing, with a well-established customer base in the corporate sector (Fig. 1).

This bank had a track record of financing significant infrastructure projects, providing crucial support for economic development. Its deep understanding of the

Fig. 1 Visual depiction of the merger of two entities (*Source* Author's creation)

corporate sector allowed it to cater to large businesses and government projects' complex financial needs.

In contrast, the retail finance leader had a robust presence in the retail segment, offering a wide range of products and services to individual consumers and small businesses. This bank was known for its customer-centric approach, providing tailored financial solutions that met the diverse needs of its retail customers. Its product portfolio included personal loans, mortgages, credit cards, and small business loans, all designed to support the financial aspirations of individuals and entrepreneurs. The retail finance leader's extensive network and customer base in the retail segment complemented the corporate bank's strengths, creating a balanced and comprehensive financial institution (Finology Insider, 2022).

The new bank aimed to leverage data analytics and advanced technology to serve a diverse customer base, from small entrepreneurs to individual consumers. This merger set the stage for a comprehensive data-driven strategy focused on delivering personalized, efficient banking services. The bank's leadership recognized the transformative potential of data analytics in driving growth and enhancing customer satisfaction. They understood that integrating data analytics into their operations could provide valuable insights into customer behaviour and preferences, enabling them to offer more relevant and personalized services.

3.2 Methodology

We accessed information from publicly available secondary sources, including industry reports, market research, financial statements, and press releases. These sources were critical in understanding the bank's data-driven initiatives and their impact on customer experience and operational efficiency.

3.2.1 Industry Reports and Market Research

We reviewed various industry reports and market research documents that provided insights into the trends and best practices in data analytics within the banking sector. These reports offered comparative analysis and benchmarks, helping to contextualize the bank's performance within the broader financial services industry.

3.2.2 Financial Statements and Annual Reports

Publicly available financial statements and annual reports were essential in detailing the bank's financial performance and strategic initiatives. These documents provided concrete evidence of the bank's revenue growth, cost savings, and operational efficiencies attributed to its data-driven strategies.

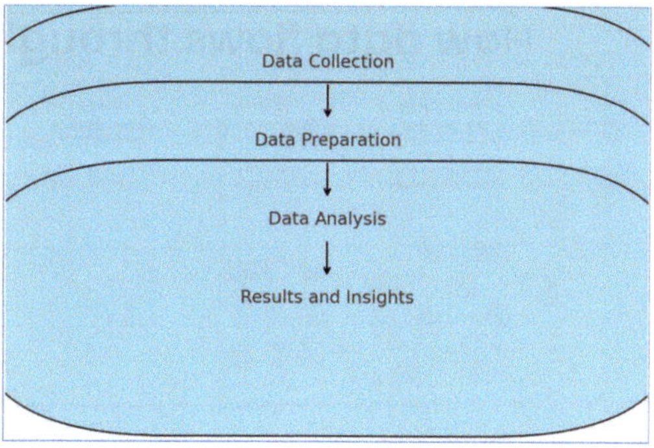

Fig. 2 Methodology (*Source* Author's creation)

3.2.3 Press Releases and News Articles

Press releases and news articles highlighted significant achievements, new product launches, and strategic partnerships of the bank. These sources provided a narrative of the bank's progress and market positioning, showcasing its commitment to data-driven decision-making and its impact on customer satisfaction and retention.

The author had the opportunity to work as an employee in their senior management cadre. At the same time, the bank was being set up, providing the author with first-hand experience of being part of and witnessing these discussions and decisions towards enhancing customer experience.

This approach ensured that our study was grounded in reliable and verifiable information, providing a robust understanding of the bank's journey and its successes in leveraging data analytics (IDFC FIRST Bank, 2020).

A visual representation of the research methodology is provided in Fig. 2 to enhance clarity and transparency:

- Data Collection: Secondary sources, including industry reports, annual statements, and press releases, were analysed.
- Data Preparation: Information was categorized into customer segmentation, operational efficiency, and risk management themes.
- Data Analysis: Analytical methods included predictive modelling, segmentation clustering, and customer feedback sentiment analysis.

4 Strategic Adoption of Data Analytics

From its inception, the new age bank placed data analytics at the core of its strategy. Recognizing the transformative potential of data, the bank invested substantially in cutting-edge technologies and bespoke analytics platforms. This section details the strategic steps to integrate data analytics into the bank's operations.

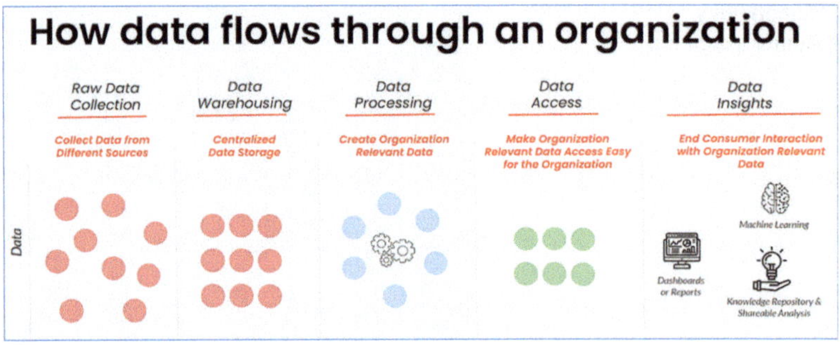

Fig. 3 Integrated data infrastructure (Reproduced from Datacamp.com) (*Source* https://www.datacamp.com/blog/what-mature-data-infrastructure-looks-like)

4.1 Developing an Integrated Data Infrastructure

A critical component of the bank's strategy was creating an integrated data infrastructure (Fig. 3). This system connected all customer-facing channels with backend systems, enabling a unified view of customer interactions. The bank embraced omnichannel banking principles, ensuring seamless access to services through multiple platforms, including mobile apps, online banking, and physical branches. The bank's data infrastructure was designed with scalability in mind, employing advanced data processing technologies such as big data frameworks and real-time analytics engines.

This setup facilitated efficient data processing and analysis, providing a robust foundation for the bank's data-driven initiatives. This integration allowed for a comprehensive view of customer interactions and behaviours, enabling the bank to offer more personalized services. Moreover, the infrastructure incorporated sophisticated data storage solutions that ensured data integrity and security.

4.2 Implementing Data-Driven Decision-Making

The bank's leadership understood that effective data utilization could unlock significant value, enabling personalized services, predictive customer insights, and enhanced operational efficiency. To achieve this, the bank developed a centralized data warehouse that aggregated data from various sources, providing a single source of truth for comprehensive data analyses. This centralized data warehouse was designed to support various data types, including structured data from transaction records and unstructured data from customer interactions on social media.

The warehouse's architecture allowed for integrating disparate data sources, facilitating a holistic analysis of customer behaviour and preferences. The data warehouse

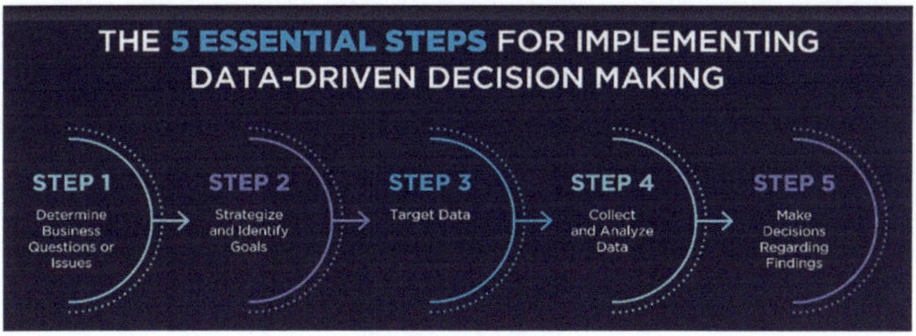

Fig. 4 Data-driven decision making (Reproduced from Jelvix.com) (*Source* https://jelvix.com/blog/data-driven-decision-making)

Fig. 5 Data analytics platform (Reproduced from freepix.com) (*Source* https://www.vectorstock.com/royalty-free-vector/data-analytics-platform-isometric-vector-23975676)

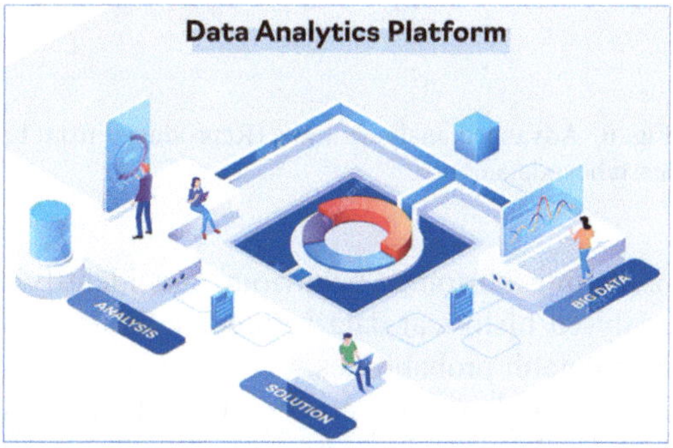

also supported real-time data processing, enabling the bank to make timely and informed decisions.

The bank implemented robust data governance practices to ensure the effectiveness of data-driven decision-making (Fig. 4). These practices included data quality management, data lineage tracking, and the establishment of data stewardship roles. The bank also invested in advanced data analytics platforms (Fig. 5) that could perform complex data analyses and generate actionable insights.

These platforms utilized machine learning algorithms to identify patterns and trends in the data, providing the bank with predictive insights that informed strategic decision-making. For instance, predictive analytics enabled the bank to anticipate customer needs and preferences, allowing it to tailor its services accordingly.

4.3 Leveraging Advanced Analytics Tools

The bank invested in advanced analytics tools (Fig. 6) to support its data-driven decision-making. These tools included machine learning algorithms for predictive analytics, natural language processing for customer sentiment analysis, and real-time data visualization platforms. Machine learning algorithms were precious in

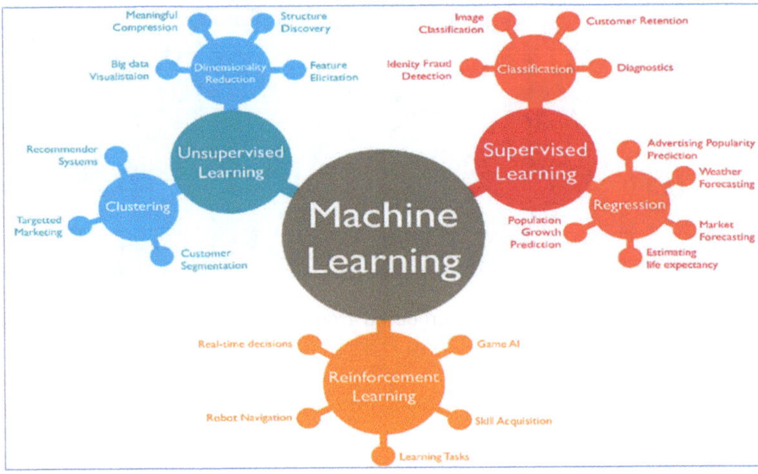

Fig. 6 Advanced analytics tools (Reproduced from hevodata.com) (*Source* https://hevodata.com/learn/hevodata.com)

predicting customer behaviour and identifying potential risks. These algorithms analysed historical data to predict future trends, such as customer churn rates or loan default probabilities.

Natural language processing (NLP) tools analysed customer feedback from various channels, including social media, emails, and call centre transcripts.

Real-time data visualization platforms provided the bank's leadership with intuitive dashboards that displayed vital performance indicators and analytics insights. These dashboards allowed for real-time business performance monitoring, enabling the bank to make data-driven decisions swiftly. The visualizations also facilitated better communication of insights across the organization, ensuring that all stakeholders were aligned with the bank's strategic objectives.

5 Data-Driven Initiatives

This section examines the specific data-driven initiatives implemented by the bank to enhance customer experience and operational efficiency.

5.1 Personalized Offerings and Proactive Engagement

The bank's focus on personalized offerings involved leveraging advanced segmentation techniques to categorize the customer base into distinct cohorts. Predictive analytics played a significant role in anticipating customer needs by analysing browsing patterns, transaction histories, and social media interactions. The bank's use of social media analytics for targeted communication further exemplifies its proactive engagement strategy (Ferm & Thaichon, 2021; Outlook Business, 2023).

For instance, the bank could proactively offer customized loan products with competitive rates by identifying customers interested in home loans through their browsing behaviour. This proactive approach improved customer satisfaction and increased the bank's loan origination volumes. The bank also utilized data from credit scores, income levels, and spending patterns to tailor loan products to individual needs, ensuring that each customer received the most suitable financial solutions.

The bank's use of chatbots powered by natural language processing (NLP) (Fig. 7) enabled it to provide personalized assistance to customers, answering queries and recommending products based on individual profiles. These chatbots could handle various customer inquiries, from account balance checks to complex loan queries, providing instant and accurate responses.

The bank employed personalized marketing campaigns (Fig. 8) on social media platforms. The bank could identify potential customers and engage them with relevant content by analysing social media interactions and user behaviour. For example, if a customer expressed interest in saving for a vacation, the bank could offer travel savings accounts or credit card offers with travel rewards.

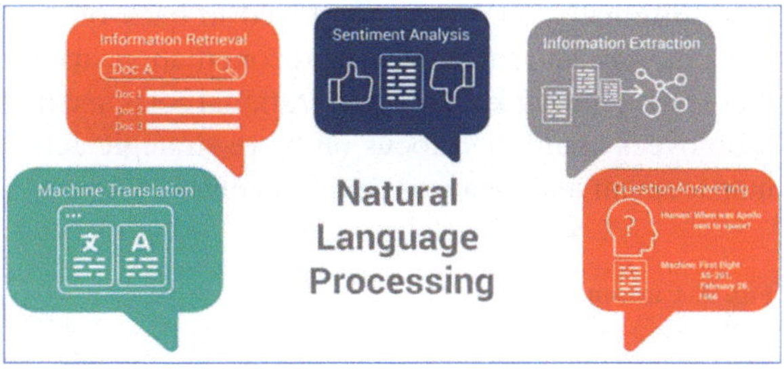

Fig. 7 Natural language processing (Reproduced from smartb.co) (*Source* https://www.smartb.co/what-is-natural-language-processing/)

Fig. 8 Personalized marketing campaigns (Reproduced from expressanalytics.com) (*Source* https://www.expressanalytics.com/solutions/customer-data-platform/marketing-analytics/)

5.2 Enhancing Operational Efficiency

Beyond customer engagement, data analytics also played a pivotal role in enhancing the bank's operational efficiency. Implementing a robust CRM system (Fig. 9) integrated with the contact centre enabled customer service representatives to have access to real-time data. This system allowed for efficient lead management, prioritizing leads based on data-driven insights, and improving conversion rates.

For instance, when a customer calls the contact centre, the CRM system provides the representative with a comprehensive view of the customer's profile, including recent interactions, transaction history, and potential product interests. This process enabled the representative to offer personalized assistance and relevant product recommendations, enhancing the overall customer experience (Outlook Business, 2024).

The bank's investment in digital operational techniques, including robotic process automation (RPA) (Fig. 10), contributed to increased efficiency by automating routine processes and streamlining workflows. For instance, RPA was used to automate the processing of loan applications, significantly reducing processing time and minimizing errors. Automated workflows ensured that applications were processed quickly and accurately, allowing customers to receive decisions faster and improving overall satisfaction.

Additionally, RPA was employed in back-office operations to handle repetitive tasks such as data entry, account reconciliation, and report generation. By automating these tasks, the bank reduced operational costs and minimized the risk of human error. Employees could then focus on more strategic activities like customer relationship management and business development.

Fig. 9 Customer relationship management (Reproduced from infetech.com) (*Source* https://www.infetech.com/crm-charts/)

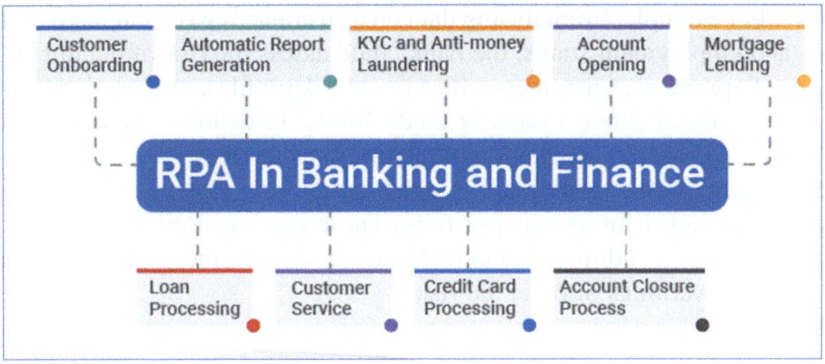

Fig. 10 Use of RPA in banking (Reproduced from xenonstack.com) (*Source* https://www.xenonstack.com/blog/rpa-for-financial-services)

5.3 Fraud Detection and Risk Management

One of the standout data analytics applications was fraud detection and risk management. The bank employed machine learning models to analyse transaction patterns and identify anomalies that could indicate fraudulent activity. For example, if a customer's account showed an unusual transaction pattern, such as a sudden large withdrawal in a foreign country, the system would flag it for further investigation. The bank's fraud detection (Fig. 11) team could then quickly assess the situation and take appropriate action, such as contacting the customer or temporarily freezing the account.

Furthermore, the bank utilized predictive analytics to assess credit risk, allowing it to make more informed lending decisions. This proactive approach reduced the bank's exposure to bad debt and enabled it to extend credit to a broader range of

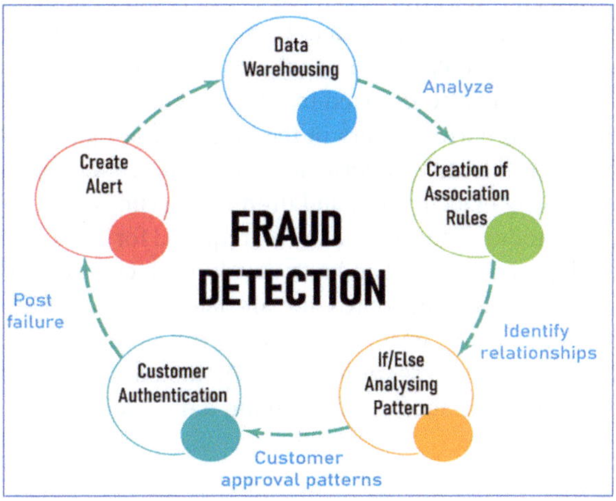

Fig. 11 Fraud detection process (Reproduced from mavink.com) (*Source* https://www.heavy.ai/technical-glossary/fraud-detection-and-prevention)

customers. For example, by analysing data on a customer's payment history, income stability, and employment status, the bank could assess the risk of default and offer tailored credit products. Customers with a lower risk profile could receive favourable terms, such as lower interest rates or credit limits. In contrast, higher-risk people might be offered secured loans or credit counselling services.

The bank also implemented real-time risk monitoring systems that continuously assessed the risk levels of its loan portfolio. These systems used advanced analytics to track economic conditions, market trends, and customer behaviour changes, providing early warnings of potential risks.

6 Addressing Challenges in Data Analytics Implementation

Despite the numerous benefits, the bank faced several challenges while implementing its data analytics strategy. This section explores these challenges and the measures taken to address them.

6.1 Ensuring Data Quality and Consistency

One of the primary challenges was ensuring data quality and consistency across multiple sources. The integration of legacy systems from both merging entities required meticulous data cleansing and standardization efforts. Disparate data formats, varying levels of data completeness, and inconsistent data entry protocols posed significant obstacles to achieving a unified data environment. To tackle these issues, the bank established a dedicated data governance team responsible for overseeing data quality initiatives. This team was tasked with implementing standardized data entry protocols to ensure uniformity in data collection and entry across all departments. They conducted regular audits to identify and rectify data discrepancies, ensuring the data used in analytics was accurate and reliable (Fig. 12).

The bank also invested in advanced data cleansing tools that could automatically detect and correct errors in the data. These tools employed machine learning algorithms to identify patterns and anomalies, facilitating removing duplicate records, correcting inaccuracies, and filling in missing data. This automated approach significantly reduced the manual effort required for data cleansing and improved the overall quality of the data. The bank developed a comprehensive data dictionary with clear definitions and standards for all data elements. This dictionary was an employee reference guide, ensuring everyone followed the same data standards and practices. Regular training sessions were conducted to familiarize staff with these standards and the importance of maintaining data quality.

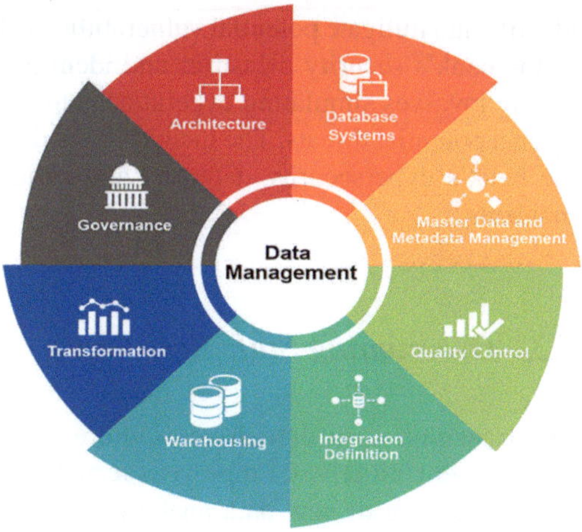

Fig. 12 Data management framework (Reproduced from 360logica.com) (*Source* https://www.360logica.com/blog/empower-software-testing-data-management/)

6.2 *Data Privacy and Security*

Data privacy and security were major concerns, especially given the sensitive nature of financial data. The bank had to navigate stringent regulatory requirements to ensure that customer data was protected and used ethically. Implementing robust data governance policies and investing in advanced cybersecurity measures were critical to addressing these concerns. To safeguard data privacy, the bank adopted encryption technologies to secure data in transit and at rest. This process ensured that data remained protected from unauthorized access during network transmission and stored in databases. Multi-factor authentication (Fig. 13) and strict access controls were implemented to prevent unauthorized access to sensitive data.

These measures ensured that only authorized personnel could access and manage critical data assets. Regular security audits and penetration testing were conducted to

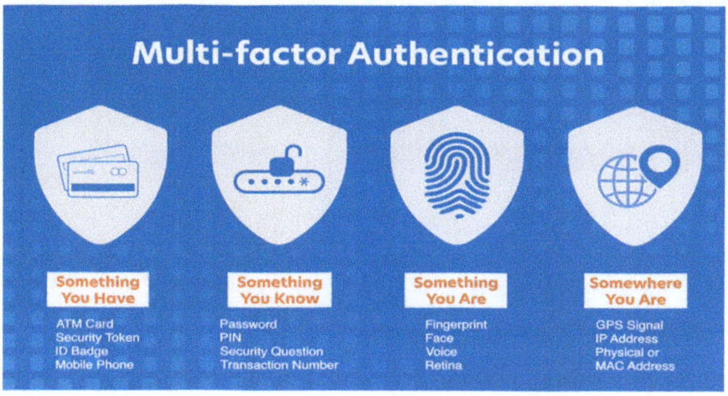

Fig. 13 Multi-factor authentication (Reproduced from ertechpros.com) (*Source* https://www.ertechpros.com/what-is-mfa-and-how-can-it-protect-your-practice)

identify and mitigate potential vulnerabilities. These audits assessed the effectiveness of the bank's security measures and identified areas for improvement. Penetration testing involves simulating cyberattacks to test the resilience of the bank's systems against potential threats. The insights gained from these tests were used to strengthen the bank's cybersecurity defences and enhance its ability to respond to security incidents.

6.3 Overcoming Staff Resistance

The transition to a data-driven culture required significant training and change management initiatives. Employees accustomed to traditional working methods often found adapting to new data analytics tools and practices challenging. To address this, the bank conducted extensive training programs to equip employees with the necessary skills and knowledge to leverage data analytics effectively. These training programs included workshops on data literacy, which helped employees understand the importance of data and how to interpret it accurately. Hands-on training sessions provided practical experience with analytics tools, enabling employees to become proficient in using them for daily tasks.

To foster a culture of data-driven decision-making, the bank implemented incentives for employees who actively engaged with data analytics. Performance metrics (Fig. 14) were adjusted to include data-driven initiatives, rewarding staff for contributing to the bank's analytics goals. Employees who demonstrated excellence in data analytics were recognized and rewarded, reinforcing the importance of data-driven practices and encouraging others to follow suit.

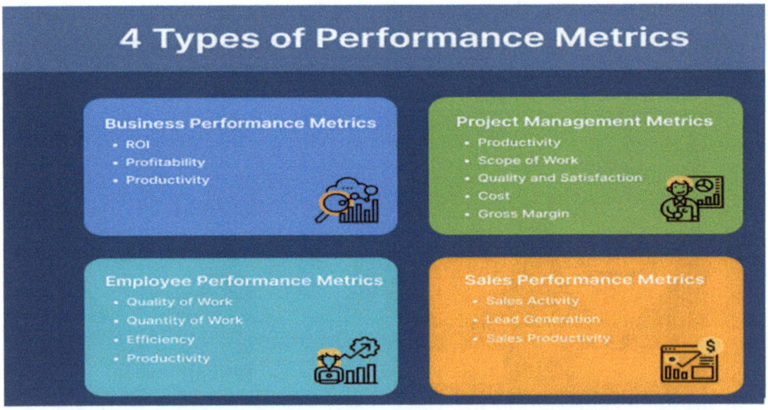

Fig. 14 Employee performance metrics (Reproduced from profit.co) (*Source* https://www.profit.co/blog/kpi-library/introductory-guide-to-understanding-performance-metrics/)

Moreover, the bank's leadership was crucial in driving the cultural shift towards data analytics. The bank also established cross-functional teams to facilitate department collaboration and ensure data analytics initiatives aligned with business objectives. These teams brought together employees from different bank areas to work on data projects, fostering a sense of ownership and accountability. The bank built a more inclusive and collaborative data-driven culture by involving employees in developing and implementing data analytics initiatives.

7 Results and Impact

The implementation of data analytics yielded tangible results for the bank. Quantitative analysis revealed significant improvements in key performance metrics, including customer satisfaction scores, retention rates, and revenue growth. Operational efficiency also improved significantly, driven by the bank's investment in digital technologies and process automation.

This section details the impact of these data-driven initiatives on various aspects of the bank's operations and performance (Fig. 15).

7.1 *Customer Satisfaction and Retention*

Customer satisfaction is a critical metric for any financial institution, and the bank's data-driven approach led to notable improvements in this area. Customer satisfaction

Fig. 15 Business performance dashboards (Reproduced from vrogue.co) (*Source* https://www.vrogue.co/post/key-performance-indicators-definition-and-examples)

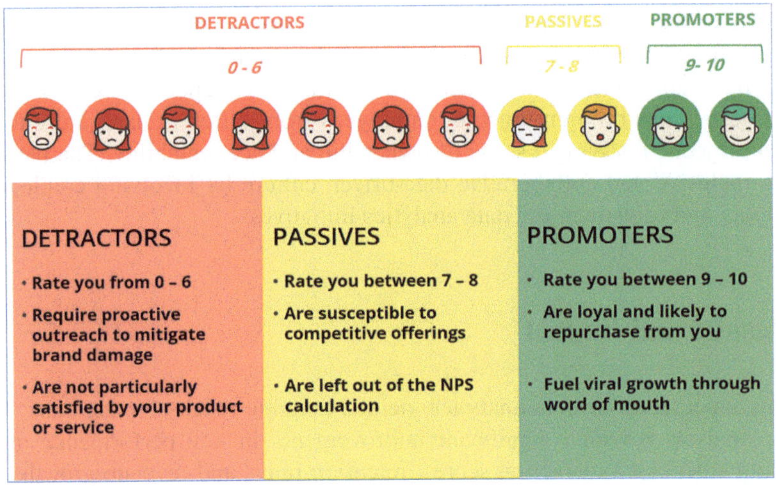

Fig. 16 Measurement of NPS (Reproduced from hairpernews.com) (*Source* https://haipernews.com/posts/how-to-measure-customer-nps/)

scores, measured through indicators like the Net Promoter Score (NPS) (Fig. 16) and Customer Satisfaction Index (CSI), showed positive trends. The bank's ability to analyse and respond to customer needs in real time was crucial in driving these improvements. Personalized offerings and proactive engagement initiatives played a significant role in enhancing customer experiences and fostering loyalty.

For example, the bank's sending personalized financial advice and product recommendations helped customers feel more valued and understood, leading to higher retention rates and customer loyalty. The tailored advice resonated well with customers, who appreciated the relevance and timeliness of the information provided. Customer feedback collected through surveys and social media indicated high satisfaction with the personalized services. The bank utilized natural language processing (NLP) tools to analyse customer sentiments and identify areas for improvement. This continuous feedback loop allowed the bank to refine its services and address any issues promptly, further enhancing customer satisfaction.

7.2 Financial Performance

From a financial perspective, the bank experienced substantial revenue growth, driven by cross-selling opportunities and wallet share expansion. By leveraging advanced predictive analytics (Fig. 17) to identify upsell and cross-sell opportunities, the bank maximized revenue generation from existing customers while attracting new business through targeted marketing campaigns and personalized promotions. The bank's focus on personalized loan products also contributed to revenue growth. The bank

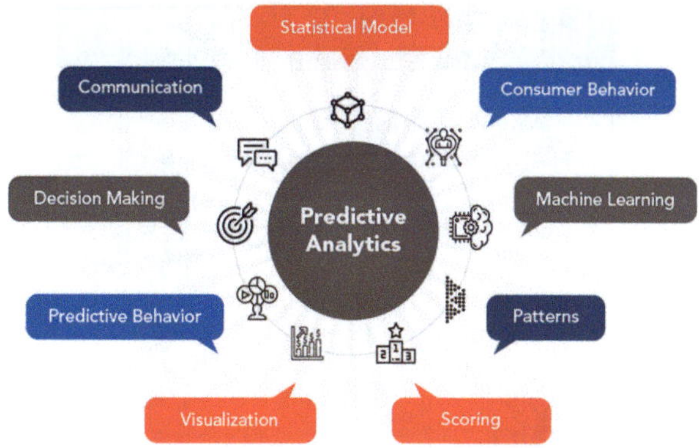

Fig. 17 Advanced predictive analytics tools (Reproduced from devteam.com) (*Source* https://www.devteam.space/blog/what-is-predictive-analysis/)

saw a significant increase in loan origination volumes by offering competitive rates and tailored loan terms based on individual customer profiles. For example, the merger of a corporate and infrastructure financing bank with a retail finance leader allowed it to diversify its loan book and increase margins, leading to a strong focus on retail loans, accounting for 63% of its loan book post-merger. Predictive models helped the bank identify customers likely to need new loans or refinancing options, enabling timely and relevant offers that met customers' needs (Lukose & Sampath, 2019).

For instance, if a customer showed a pattern of increased spending on home improvement items, the bank could proactively offer a home equity loan or a refinancing option. This targeted approach met the customers' immediate financial needs and increased the bank's loan portfolio and revenue.

The bank's data-driven marketing campaigns also played a significant role in attracting new customers. The bank could create highly targeted marketing campaigns that resonated with specific customer segments by analysing demographic data, social media interactions, and purchasing behaviours. These campaigns resulted in higher conversion rates and increased market penetration. The bank's identification and mitigation of credit risks (Fig. 18) using predictive analytics helped maintain a healthy loan portfolio.

By assessing the creditworthiness of potential borrowers more accurately, the bank reduced the incidence of non-performing loans, contributing to overall financial stability and profitability. The bank's financials show an improvement in net interest margin (NIM) to 5.09% post-merger, driven by the increased focus on high-yield retail loans and a higher CASA ratio, which stood at 51.75%.

7.3 Operational Efficiency

The bank's data-driven initiatives improved operational efficiency (Fig. 19) significantly. The bank's investment in digital technologies and process automation led

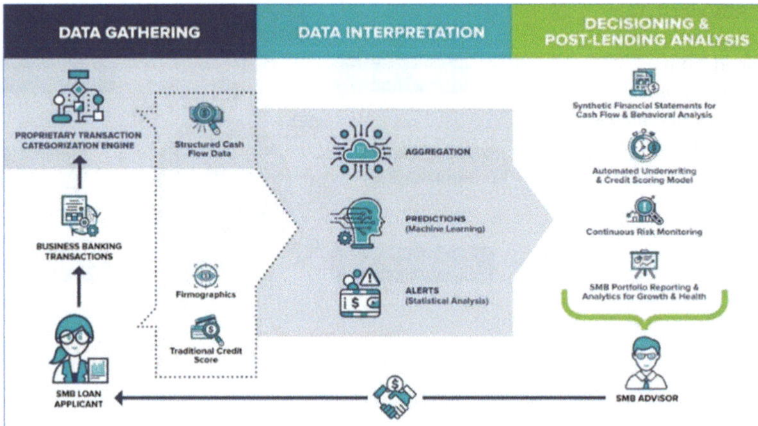

Fig. 18 Credit underwriting tools (Reproduced from judi.ai) (*Source* https://judi.ai/underwriting-science/)

to substantial cost savings and enhanced regulatory compliance. By proactively detecting and mitigating fraudulent activities, the bank minimized financial losses and maintained the integrity of its banking system.

Implementing robotic process automation (RPA) for routine tasks such as data entry and transaction processing significantly reduced operational overheads. RPA bots could quickly and accurately handle high-volume, repetitive tasks, freeing staff to focus on more strategic activities. This not only reduced operational costs but also improved process efficiency and accuracy. For example, the bank's operational expenses declined, as indicated by reduced cost-to-income ratios over recent years. Additionally, the streamlined loan processing workflow reduced approval times by a significant margin, enhancing customer satisfaction.

Fig. 19 Process efficiency improvement (Reproduced from paperfreecorp.com) (*Source* https://paperfreecorp.com/solutions/information-capture)

Automated systems could quickly verify application details, check credit scores, and process loan approvals, reducing the time customers had to wait for loan decisions. This speed and efficiency were particularly beneficial for small businesses and consumers needing quick access to funds. The bank's focus on efficiency improvements can be seen in its operational metrics, where automation and digitalization have led to faster service delivery and improved customer experiences. The bank's use of advanced analytics (Fig. 20) for performance monitoring and optimization also played a crucial role in improving operational efficiency. Real-time data dashboards gave managers insights into key performance metrics, enabling them to identify bottlenecks and optimize processes. The integration of real-time analytics allowed the bank to continuously monitor and enhance its service delivery processes, ensuring high customer service standards.

Integrating AI and machine learning models in fraud detection systems significantly enhanced the bank's ability to detect and prevent fraudulent activities. These models could analyse vast amounts of transaction data in real time, identifying unusual patterns that might indicate fraud. Early detection of fraudulent activities helped the bank minimize financial losses and protect its customers' assets. The proactive approach to fraud prevention, supported by advanced analytics, has been pivotal in maintaining the bank's reputation and financial stability.

These systems provided a comprehensive view of the bank's risk exposure, enabling proactive risk mitigation strategies. The bank maintained a robust risk management framework and ensured long-term stability and resilience. The focus on data-driven risk management practices has allowed the bank to anticipate better and mitigate potential risks, contributing to overall financial health.

Research Contribution and Novelty

This study contributes to the field in the following ways:

Fig. 20 Advanced risk management framework (Reproduced from vrogue.co) (*Source* https://www.vrogue.co/post/the-risk-analysis-framework-with-the-elements-risk-assessment-risk)

1. **Addressing Research Gaps:** While previous studies have focused on data analytics in advanced economies, this study highlights its practical application in India, a developing market with unique challenges.
2. **Novel Insights:** The study introduces a data-driven approach to customer experience enhancement, demonstrating how clustering, predictive modelling, and sentiment analysis can be integrated into banking processes.
3. **Theoretical Contribution:** By extending the concepts of Customer Journey Mapping and CRM, this study provides a framework for analytics adoption in resource-constrained environments.

Practical Implications

The findings of this study provide actionable insights for banking practitioners. For example:

- **Personalized Services:** Banks can use data analytics to design tailored financial products and services, increasing customer satisfaction and loyalty.
- **Operational Efficiency:** Automating processes like loan approvals and fraud detection through advanced analytics reduces operational costs and improves turnaround times.
- **Risk Management:** Predictive modelling helps banks identify potential defaulters and mitigate risks proactively.

These practices can benefit banks in emerging markets struggling with legacy systems and intense competition from fintech startups.

Theoretical Implications

This study contributes to the theoretical understanding of data analytics in banking by:

- Extending the application of Customer Journey Mapping to include real-time data-driven decision-making across touchpoints.
- Reinforcing the role of CRM frameworks in leveraging customer insights for targeted engagement.
- Highlighting the scalability of analytics solutions in emerging markets, providing a basis for future research on technology adoption in resource-constrained environments.

8 Limitations and Future Research Agenda

This study has mainly focused on the impact of data analytics on enhancing customer experience in the financial service sector. Future studies can utilise the insights and findings from this study to gauge the impact of data analytics on enhancing customer experience in other industry contexts. This study is about a new-age bank in an established industry context; hence, it may not reflect challenges and constraints posed by legacy systems and processes that may mar similar technological investments toward improving customer experience. Future studies can investigate this aspect of new vs legacy businesses and their returns on investment (ROI) from such strategic decisions. Emerging technologies such as artificial intelligence (AI), machine learning (ML), blockchain, and quantum computing promise to unlock new frontiers in data analytics and customer experience enhancement.

Declaration of Originality We, Rupali, Shilpa, and Dinesh, hereby declare that the content of the book chapter titled "***Towards Customer Experience Enhancement using Customer Analytics: A Case Study of a Leading Indian Private Bank***" is entirely original and has been written solely by us. We confirm that we have selectively used AI to create an illustration in Fig. 1 to visually support our commentary, facilitating a better understanding of the case study.

We further affirm that all sources and references have been appropriately cited, and the chapter adheres to the publication's guidelines on originality and ethical writing practices.

References

Al-Dmour, H., Saad, N., Basheer Amin, E., Al-Dmour, R., & Al-Dmour, A. (2023). The influence of the practices of big data analytics applications on bank performance: Filed study. *VINE Journal of Information and Knowledge Management Systems, 53*(1), 119–141.

Alnawas, I., & Hemsley-Brown, J. (2019). Examining the key dimensions of customer experience quality in the hotel industry. *Journal of Hospitality Marketing & Management, 28*(7), 833–861.

Ascarza, E., Ebbes, P., Netzer, O., & Danielson, M. (2017). Beyond the target customer: Social effects of customer relationship management campaigns. *Journal of Marketing Research, 54*(3), 347–363.

Becker, J. U., Greve, G., & Albers, S. (2009). The impact of technological and organizational implementation of CRM on customer acquisition, maintenance, and retention. *International Journal of Research in Marketing, 26*(3), 207–215.

Becker, L., & Jaakkola, E. (2020). Customer experience: Fundamental premises and implications for research. *Journal of the Academy of Marketing Science, 48*, 630–648.

Bolton, R. N., McColl-Kennedy, J. R., Cheung, L., Gallan, A., Orsingher, C., Witell, L., & Zaki, M. (2018). Customer experience challenges: Bringing together digital, physical and social realms. *Journal of Service Management, 29*(5), 776–808.

Brakus, J. J., Schmitt, B. H., & Zarantonello, L. (2009). Brand experience: What is it? How is it measured? Does it affect loyalty? *Journal of Marketing, 73*(3), 52–68.

Bueno, E. V., Weber, T. B. B., Bomfim, E. L., & Kato, H. T. (2019). Measuring customer experience in service: A systematic review. *The Service Industries Journal, 39*(11–12), 779–798.

Chahal, H., & Dutta, K. (2015). Measurement and impact of customer experience in banking sector. *Decision, 42*, 57–70.

Chang, W., Park, J. E., & Chaiy, S. (2010). How does CRM technology transform into organizational performance? A mediating role of marketing capability. *Journal of Business Research, 63*(8), 849–855.

Chauhan, S., Akhtar, A., & Gupta, A. (2022). Customer experience in digital banking: A review and future research directions. *International Journal of Quality and Service Sciences, 14*(2), 311–348.

Cho, J., DeStefano, T., Kim, H., Kim, I., & Paik, J. H. (2023). What's driving the diffusion of next-generation digital technologies? *Technovation, 119*, Article 102477.

Côrte-Real, N., Ruivo, P., Oliveira, T., & Popovič, A. (2019). Unlocking the drivers of big data analytics value in firms. *Journal of Business Research, 97*, 160–173.

Davenport, T. H., & Westerman, G. (2018). Why so many high-profile digital transformations fail. *Harvard Business Review, 9*(4), 15.

De Keyser, A., Lemon, K. N., Klaus, P., & Keiningham, T. L. (2015). A framework for understanding and managing the customer experience. Marketing Science Institute working paper series, 85(1), 15–121.

Ferm, L. E. C., & Thaichon, P. (2021). Value co-creation and social media: Investigating antecedents and influencing factors in the US retail banking industry. *Journal of Retailing and Consumer Services, 61*, Article 102548.

Finology Insider. (2022). https://insider.finology.in/business/idfc-first-bank-business-model

Frizzo-Barker, J., Chow-White, P. A., Mozafari, M., & Ha, D. (2016). An empirical study of the rise of big data in business scholarship. *International Journal of Information Management, 36*(3), 403–413.

Frow, P., & Payne, A. (2007). Towards the 'perfect'customer experience. *Journal of Brand Management, 15*(2), 89–101.

Gambetta, N., García-Benau, M. A., & Zorio-Grima, A. (2016). Data analytics in banks' audit: The case of loan loss provisions in Uruguay. *Journal of Business Research, 69*(11), 4793–4797.

Grewal, D., Levy, M., & Kumar, V. (2009). Customer experience management in retailing: An organizing framework. *Journal of Retailing, 85*(1), 1–14.

Groth, M., Wu, Y., Nguyen, H., & Johnson, A. (2019). The moment of truth: A review, synthesis, and research agenda for the customer service experience. *Annual Review of Organizational Psychology and Organizational Behavior, 6*, 89–113.

Gurbaxani, V., & Dunkle, D. (2019). Gearing up for successful digital transformation. *MIS Q. Executive, 18*(3), 6.

Gustafsson, A., Caruelle, D., & Bowen, D. E. (2024). Customer experience (CX), employee experience (EX) and human experience (HX): Introductions, interactions and interdisciplinary implications. *Journal of Service Management, 35*(3), 333–356. https://doi.org/10.1108/JOSM-02-2024-0072

Hajiheydari, N., Delgosha, M. S., Wang, Y., & Olya, H. (2021). Exploring the paths to big data analytics implementation success in banking and financial service: An integrated approach. *Industrial Management & Data Systems, 121*(12), 2498–2529.

Hillebrand, B., Nijholt, J. J., & Nijssen, E. J. (2011). Exploring CRM effectiveness: An institutional theory perspective. *Journal of the Academy of Marketing Science, 39*, 592–608.

Holbrook, M. B., & Hirschman, E. C. (1982). The experiential aspects of consumption: Consumer fantasies, feelings, and fun. *Journal of Consumer Research, 9*(2), 132–140.

Holmlund, M., Van Vaerenbergh, Y., Ciuchita, R., Ravald, A., Sarantopoulos, P., Ordenes, F. V., & Zaki, M. (2020). Customer experience management in the age of big data analytics: A strategic framework. *Journal of Business Research, 116*, 356–365.

How can banks thrive in a digital society? (2021). *Deloitte Netherlands*. https://www2.deloitte.com/nl/nl/pages/financial-services/articles/tech-trends-banking.html

Hyper-personalization: the next phase of banks' digital evolution - Wipro. (2021, May). https://www.wipro.com/blogs/mahesh-g-raja/hyper-personalization-the-next-phase-of-banks-digital-evolution/

IDFC FIRST Bank. (2020). *Corporate Presentation—Q1 FY20*. https://www.idfcfirstbank.com/content/dam/idfcfirstbank/pdf/financial-results/IDFCFIRSTB-Investor-Presentation-Commentary-Q1FY20.pdf

Jain, R., Aagja, J., & Bagdare, S. (2017). Customer experience – A review and research agenda. *Journal of Service Theory and Practice, 27*(3), 642–662. https://doi.org/10.1108/jstp-03-2015-0064

Kandampully, J., Zhang, T. C., & Jaakkola, E. (2018). Customer experience management in hospitality: A literature synthesis, new understanding and research agenda. *International Journal of Contemporary Hospitality Management, 30*(1), 21–56.

Klink, R. R., Zhang, J. Q., & Athaide, G. A. (2020). Measuring customer experience management and its impact on financial performance. *European Journal of Marketing, 55*(3), 840–867.

Kranzbühler, A. M., Kleijnen, M. H., Morgan, R. E., & Teerling, M. (2018). The multilevel nature of customer experience research: An integrative review and research agenda. *International Journal of Management Reviews, 20*(2), 433–456.

Kuehnl, C., Jozic, D., & Homburg, C. (2019). Effective customer journey design: Consumers' conception, measurement, and consequences. *Journal of the Academy of Marketing Science, 47*, 551–568.

Kumar, P. (2022). Managing service flexibility in healthcare for improved customer experience: a data-driven approach. Journal of Strategic Marketing, 1–22.

Kumari, M. (2019). Effect of 'Demonetization 2016 ' on Indian Economy. *Available at SSRN 3783805.*

Lehrer, C., Wieneke, A., Vom Brocke, J. A. N., Jung, R., & Seidel, S. (2018). How big data analytics enables service innovation: Materiality, affordance, and the individualization of service. *Journal of Management Information Systems, 35*(2), 424–460.

Lemon, K. N., & Verhoef, P. C. (2016). Understanding customer experience throughout the customer journey. *Journal of Marketing, 80*(6), 69–96.

Lukose, J., & Sampath, A. (2019). IDFC Bank and capital first – a case of merger arbitrage. In Indian Institute of Management, Kozhikode, *Indian Institute of Management Kozhikode*. https://iimk.ac.in/uploads/publications/3119For%20Upload%2021081_new.pdf

Maklan, S., & Klaus, P. (2011). Customer experience: Are we measuring the right things? *International Journal of Market Research, 53*(6), 771–772.

Maklan, S., Antonetti, P., & Whitty, S. (2017). A better way to manage customer experience: Lessons from the royal bank of Scotland. *California Management Review, 59*(2), 92–115.

Marinakos, G., & Daskalaki, S. (2017). Imbalanced customer classification for bank direct marketing. *Journal of Marketing Analytics, 5*, 14–30.

Mathew, S. M. (2024). Exploring the exponential growth of UPI in India: A study on digital payment transformation (2016–2024). *Library Progress International, 44*(3), 10796–10805.

McColl-Kennedy, J. R., Gustafsson, A., Jaakkola, E., Klaus, P., Radnor, Z. J., Perks, H., & Friman, M. (2015). Fresh perspectives on customer experience. *Journal of Services Marketing, 29*(6/7), 430–435.

Meyer, C., & Schwager, A. (2007). Understanding customer experience. *Harvard Business Review, 85*(2), 116.

Micheaux, A., & Bosio, B. (2019). Customer journey mapping as a new way to teach data-driven marketing as a service. *Journal of Marketing Education, 41*(2), 127–140.

Moeyersoms, J., & Martens, D. (2015). Including high-cardinality attributes in predictive models: A case study in churn prediction in the energy sector. *Decision Support Systems, 72*, 72–81.

Nguyen, A. H., Hoang, T. G., Nguyen, L. Q. T., & Thi Pham, H. M. (2024). Design thinking-based data analytic lifecycle for improving management control in banks. *Technology Analysis & Strategic Management, 36*(7), 1508–1523.

Outlook Business. (2023). *IDFC first is digitally Savvy, knows customers well to build services around their needs: Madhivanan Balakrishnan | Outlook Business*. https://www.outlookbusiness.com/. https://www.outlookbusiness.com/strategy-4/interview-18/idfc-first-is-digitally-savvy-knows-customers-well-to-build-services-around-their-needs-madhivanan-balakrishnan-6869

Outlook Business. (2024). *Obsessed with doing what is right for customers: IDFC FIRST Bank's Chinmay Dhoble | Outlook Business*. https://www.outlookbusiness.com/. https://www.outlookbusiness.com/enterprise-3/feature/obsessed-with-ng-what-is-right-for-customers-idfc-first-banks-chinmay-dhoble-7044

Patrı´Cio, L., Fisk, R. P., Falca˜O E Cunha, J., & Larry Constantine. (2011). Multilevel service design: from customer value constellation to service experience blueprinting. In Journal of Service Research (Vols. 14–2, pp. 180–200) [Journal-article]. https://doi.org/10.1177/1094670511401901

Parasuraman, A., Ball, J., Aksoy, L., Keiningham, T. L., & Zaki, M. (2020). More than a feeling? Toward a theory of customer delight. *Journal of Service Management, 32*(1), 1–26. https://doi.org/10.1108/josm-03-2019-0094

Peltier, J. W., Dahl, A. J., & Swan, E. L. (2020). Digital information flows across a B2C/C2C continuum and technological innovations in service ecosystems: A service-dominant logic perspective. *Journal of Business Research, 121*, 724–734.

"Phil" Klaus, P., & Maklan, S. (2012). EXQ: a multiple-item scale for assessing service experience. *Journal of Service Management, 23*(1), 5–33.

Pine, B. J., & Gilmore, J. H. (1998). Welcome to the experience economy. *Harvard Business Review, 76*(4), 97–106.

Pradhan Mantri Jan-Dhan Yojana (PMJDY)| National Portal of India. (n.d.). https://www.india.gov.in/spotlight/pradhan-mantri-jan-dhan-yojana-pmjdy#tab=tab-1

Rahimian, S., ShamiZanjani, M., Manian, A., & Esfidani, M. R. (2021). A framework of customer experience management for hotel industry. *International Journal of Contemporary Hospitality Management, 33*(5), 1413–1436.

Saggi, M. K., & Jain, S. (2018). A survey towards an integration of big data analytics to big insights for value-creation. *Information Processing & Management, 54*(5), 758–790.

Schmitt, B. (1999). Experiential marketing. *Journal of Marketing Management, 15*(1–3), 53–67.

Siebert, A., Gopaldas, A., Lindridge, A., & Simões, C. (2020). Customer experience journeys: Loyalty loops versus involvement spirals. *Journal of Marketing, 84*(4), 45–66.

Singh, N. (2022). Digital India as an innovation system. In *Innovation systems, economic development and public policy* (pp. 221–239). Routledge India.

Silva, J. H., Mendes, G. H., Miguel, P. A. C., Amorim, M., & Teixeira, J. G. (2021). Customer experience research: Intellectual structure and future research opportunities. *Journal of Service Theory and Practice, 31*(6), 893–931.

Singh, A. K., & Sharma, P. (2022). A study of Indian Gen X and Millennials consumers' intention to use FinTech payment services during COVID-19 pandemic. *Journal of Modelling in Management, 18*(4), 1177–1203.

Soltani Delgosha, M., Hajiheydari, N., & Fahimi, S. M. (2021). Elucidation of big data analytics in banking: A four-stage Delphi study. *Journal of Enterprise Information Management, 34*(6), 1577–1596.

Sorescu, A., & Schreier, M. (2021). Innovation in the digital economy: A broader view of its scope, antecedents, and consequences. *Journal of the Academy of Marketing Science, 49*(4), 627–631.

Soto Setzke, D., Riasanow, T., Böhm, M., & Krcmar, H. (2023). Pathways to digital service innovation: The role of digital transformation strategies in established organizations. *Information Systems Frontiers, 25*(3), 1017–1037.

Stake, R. (1995). *Case study research.* Sage.

Tueanrat, Y., Papagiannidis, S., & Alamanos, E. (2021). Going on a journey: A review of the customer journey literature. *Journal of Business Research, 125*, 336–353.

Vargo, S. L., & Lusch, R. F. (2004). Evolving to a new dominant logic for marketing. *Journal of Marketing, 68*(1), 1–17.

Vargo, S. L., Koskela-Huotari, K., & Vink, J. (2020). Service-dominant logic: foundations and applications. In *The Routledge handbook of service research insights and ideas* (pp. 3–23). Routledge.

Verhoef, P. C., Lemon, K. N., Parasuraman, A., Roggeveen, A., Tsiros, M., & Schlesinger, L. A. (2009). Customer experience creation: Determinants, dynamics and management strategies. *Journal of Retailing, 85*(1), 31–41.

Voorhees, C. M., Fombelle, P. W., Gregoire, Y., Bone, S., Gustafsson, A., Sousa, R., & Walkowiak, T. (2017). Service encounters, experiences and the customer journey: Defining the field and a call to expand our lens. *Journal of Business Research, 79*, 269–280. https://doi.org/10.1016/j.jbusres.2017.04.014

Wetzels, R. W., & Wetzels, M. (2023). There is a secret to success: Linking customer experience management practices to profitability. *Journal of Retailing and Consumer Services, 73*, Article 103338.
World Retail Banking Report 2024 | Intelligent Banking I Capgemini. (2024, June 6). *Capgemini.* https://www.capgemini.com/insights/research-library/world-retail-banking-report/
Yin, R. K. (2012). *Applications of case study research* (Vol. 34). Sage.
Yin, R. K. (2002). *Case study research: Design and methods* (3rd ed.). Sage.
Zaki, M. (2019). Digital transformation: Harnessing digital technologies for the next generation of services. *Journal of Services Marketing, 33*(4), 429–435.

Returns Management and Reverse Logistics in Omnichannel Supply Chain: A Systematic Review and Future Research Agenda from a Closed Loop Supply Chain Perspective

Anukesh Valase and Arshinder Kaur

Abstract *Purpose* Despite the numerous challenges, omnichannel retailers often overlook returns management and reverse logistics, and there is a lack of literature addressing these challenges. This systematic review summarizes the existing research in these areas, highlights diverse themes based on the decision problems within each phase of an omnichannel supply chain from a closed-loop perspective, and proposes future research directions through a conceptual framework. *Design/Methodology/Approach* We consider 22 articles published between 2016 and 2023 for the systematic review. The investigation includes categorizing papers into five themes and expanding on the consequences of various challenges associated with each theme by adopting a closed-loop perspective for an Omnichannel Supply Chain. *Findings* The study's findings indicate a scarcity of literature on returns management and reverse logistics in OC, suggesting that this subject of study is still in its infancy. The review clearly highlights the need to address the forward and reverse flows of an OC jointly to consider an Omnichannel Closed Loop Supply Chain for optimal results. The proposed framework adopts a closed-loop view of the omnichannel supply chain and outlines thematic future research directions. *Limitations* We restrict the review to publications that are indexed in the Web of Science and Scopus databases and to English-language articles. Future systematic reviews may include papers from more databases, such as ScienceDirect, ProQuest, and others, as well as publications written in other languages. *Originality/Value and Implications* This review provides practitioners with valuable insights into the key challenges of returns management and reverse logistics in omnichannel retailing, emphasizing the importance of a closed-loop perspective. Theoretically, it contributes to the literature by introducing a new closed-loop approach that integrates forward and reverse flows, and it identifies future research opportunities in this area.

A. Valase (✉) · A. Kaur
Indian Institute of Technology Madras, Chennai, India
e-mail: valaseanukesh@gmail.com

A. Kaur
e-mail: arshinder@iitm.ac.in

B. Basu et al. (eds.), *Data Analytics for Decision Making towards Business Excellence*, Uncertainty and Operations Research, https://doi.org/10.1007/978-981-95-2429-7_4

Keywords Omnichannel retailing · Returns · Reverse logistics · Systematic literature review · Omnichannel closed-loop supply chain

1 Introduction

The growth of e-commerce has been fuelled by increasing customer needs for convenience, wide assortment, and quick deliveries owing to technological advancements. Most pure brick-and-mortar retailers have added online channels to their existing brick-and-mortar channels and have extensively implemented dual and multi-channel retailing in order to enhance the value proposition and expand their reach to more customers compared to pure online or offline strategies (Zhang et al., 2010). Moreover, modern consumers are astute enough to evaluate all the accessible channels for an optimized purchase decision (Rigby, 2011). However, dual and multi-channel retailing involve parallel and independent operation of channels resulting in a mismatch of information and service inconsistencies (Beck & Rygl, 2015). Retailers have recognized the need for channel integration in order to deliver a seamless customer experience, raise the value proposition to consumers, and increase operational efficiency (Rigby, 2011). Considering the shortcomings associated with independent channels in dual and multichannel settings, retailers adopted the omnichannel strategy wherein all the available channels are integrated so that the channels are not differentiated by the retailer and the customer (Wollenburg et al., 2019).

According to (Verhoef et al., 2015), OC management is "the synergetic management of the numerous available channels and customer touchpoints, in such a way that the customer experience across channels and the performance over channels are optimized." While the definition provided by (Verhoef et al., 2015) describes OC management in general, (Taylor et al., 2019) attempt to delineate between OC and other strategies and define OC fulfillment as "Processes that enable a firm to meet customer demand through the flexible sharing of fulfillment link(s) across any combination of channels with respect to purchase origination and purchase receipt." Thus, (Taylor et al., 2019) assert that omnichannel fulfillment entails flexible sharing of personnel, inventory, and other fulfillment assets for order fulfillment irrespective of the channel of order origination and the channel of order receipt. (Melacini et al., 2018) delineate multichannel and omnichannel strategies in terms of their focus, integration, management, and data. They state that, whereas multichannel concentrates only on interactive channels, OC also considers mass communication channels. They also point out that multichannel retailing consists of separate channels with no integration, channel-specific objectives for the management of channels, and no integration/sharing of data across the available channels, whereas OC retailing emphasises on integrating available channels for a seamless retail experience, thus promoting cross-channel objectives over channel-specific objectives and integration of the data across channels. We only take into account papers that conform to these definitions of OC and OC fulfilment when selecting the sample of papers for the review, and exclude the non-conforming articles due to the discrepancies in the definition of omnichannel retailing.

Several offline retailers such as Walmart and Best Buy have integrated online channels into their existing physical store network to implement omnichannel retailing and provide a seamless experience across channels, enhance customer convenience and operational efficiency. On the other hand pure online retailers such as Amazon and Warby Parker have established offline presence to implement omnichannel retailing. Therefore, despite the growing demand for e-commerce omnichannel retailing has attracted substantial attention from retailers due to the drawbacks of e-commerce. E-commerce retailing, although it provides a wide variety of assortments (Rooderkerk & Kök, 2019) and circumvents geographical barriers (Brynjolfsson et al., 2013; Saghiri et al., 2017), it does not provide the customers with an opportunity to inspect the products physically (L. Liu et al., 2020; Mandal et al., 2021). The inability to touch and feel the product creates a barrier in communicating the information through the online channel, thus elevating the product valuation uncertainty among the customers and engendering massive returns (L. Liu et al., 2020). Consequently, retailers are confronted with the challenge of managing a glut of returns which entails monumental costs (Mandal et al., 2021). For these reasons, retailers have traditionally seen returns as a "necessary evil" and an afterthought and hence are confronted with the task of handling massive return volumes (Muir et al., 2019). Returns, on the other hand, may be used to gain a competitive advantage via well-designed reverse logistics and returns management strategies leading to enhanced customer satisfaction and fulfillment strategies (Muir et al., 2019). Additionally, online retailers experience higher returns processing costs than traditional retailers since traditional retailers utilize the physical stores for gatekeeping of the returns (Grewal et al., 2004).

However, in an integrated OC scenario, the retailers can leverage the physical stores for in-store returns, which are not only convenient for the customers but also for the retailer as the customer can be offered replacement product as an exchange, leading to cross-selling opportunities and a significant reduction in transportation costs of returned products (Wollenburg et al., 2018). In an omnichannel setting, the return of a product is not coupled with the forward channel, and the customer has the liberty to choose a different channel for return (Hübner et al., 2016). (Bell et al., 2018) in their study, observed a lower return rate in areas where an OC retailer has a physical presence. Thus omnichannel retailing can assist in efficient management of returned products. Cross-channel returns can be consolidated at the nearest physical store and added to the store inventory, thus getting the product back into the forward flow at a faster rate and reducing the returns handling costs (Agatz et al., 2008; Hübner et al., 2015). However, these factors have increased the complexity and expense of processing OC returns, highlighting the critical need for designing effective returns management and reverse logistics strategies for OC retailing (De Leeuw et al., 2016). Despite the inherent complexity of reverse logistics management owing to the many uncertainties associated with product returns and customer channel choices for return, the reverse flow in an OC setting remains under-explored to a large extent (Xu & Jackson, 2019).

Systematic reviews have been conducted in the area of e-commerce returns (Abdulla et al., 2019; Ambilkar et al., 2022; Duong et al., 2022; Govindan & Bouzon, 2018). However, as the omnichannel revolution gains pace in the retailing industry globally, there is a need to comprehend, investigate, and address the essential challenges in order to develop a streamlined and efficient system. Furthermore, the omnichannel adoption further introduces complexities in returns management and reverse logistics, necessitating a focused investigation on these evolving challenges. Several literature reviews on OC have been undertaken in the past, but the emphasis of these literature reviews have been on the areas of forward flow such as e-fulfillment and distribution (Melacini et al., 2018), OC fulfillment (Taylor et al., 2019), warehouse operations, and design (Kembro et al., 2018).

In the past, (De Borba et al., 2020) conducted an SLR on returns in OC with the aim of identifying the barriers to returns management in OC. However, there remains a significant gap in the literature: there is no comprehensive review that categorizes the existing literature on returns management and reverse logistics in OC to provide a theme-wise map based on the decision problem and the phase of the omnichannel journey. Moreover, the substantial body of recent research in this domain further underscores the need for an updated systematic literature review to synthesize these findings and advance the academic understanding as well as the managerial implications.

As a result, we identified a need to synthesize the existing body of knowledge in the form of a systematic literature review and provide the groundwork for future research in this field. This systematic review would thus contribute to the literature by highlighting the gaps in the research on returns management and reverse logistics in OC retailing. The primary objective of this paper is to identify the themes emerging from the literature to provide a comprehensive and structured assessment of the existing body of knowledge, thus providing a holistic view of the underlying literature. Additionally, it benefits researchers and practitioners by laying the groundwork for future research aimed at optimizing returns management and reverse logistics in OC retailing and for the research community to better understand and build upon the critical phenomenon. This systematic review enhances the existing findings and deepens the understanding of the subject area. The classification of the literature into themes would increase our comprehension of the present state of the literature in each theme, while also laying the groundwork for future study potential. Thus, it offers a thematically comprehensive summary of the fragmented literature on returns management and reverse logistics in OC. This study intends to determine the evolving requirements for research on many facets of returns management and reverse logistics in OC, as retailers attempt to use a multitude of returns management strategies to deal with the influx of returns.

We aim to answer the following research questions through the systematic review of the literature:

1. What are the various themes that emerge based on the decision problems in the returns management and reverse logistics in an OCSC from a Closed-loop perspective?

2. What is the current state of the literature within each theme?
3. What are the various issues to be addressed that constitute the future research prospects within each theme with respect to addressing the OCSC from a CLSC perspective?

2 Scope of the Review and Framework Development

Considering the gap in the literature, we focus only on those articles that address returns management or reverse logistics in OC retailing and exclude the rest. Although reverse logistics and returns management in dual/multi-channel scenarios have been addressed in the literature, there is a dearth of such studies in an OC scenario. Despite the fact that the phrases dual channel, multi-channel and omnichannel have been used interchangeably, we only evaluate articles that comply to the omnichannel definition stated in Sect. 1 of the article and omit those that do not. Omnichannel returns management differs significantly from returns management for e-commerce, dual-channel, and multi-channel retailing. In OCR, strategies such as showrooming are adopted to curb returns and BORIS to reduce reverse logistics costs and enhance customer convenience (Hao et al., 2024). The integration of multiple channels in an omnichannel context adds complexity to the returns management process, introducing uncertainties in quantity, quality and condition of returns across several channels. Retailers thus need to carefully consider inventory decisions and return policies as these strategies have a significant impact on profitability and inventory levels. Effective management of returns in an omnichannel context requires sophisticated logistics, information systems, and advanced technologies to handle the increased complexity and ensure seamless customer experience (Bernon et al., 2016).

In this section, we aim to propose a framework to classify the literature on returns management and reverse logistics in OC retailing. We categorize the various processes into three phases while considering the reverse flow. Given that channels are treated independently in a dual/multi-channel setting as opposed to an integrated approach in an omnichannel scenario, we limit the scope of this review to articles addressing OC retailing that adhere to the definition provided in Sect. 1 of this article, which clearly differentiates dual/multichannel retailing from omnichannel retailing.

OC retailing is a significant driver that propels innovation in inventory, procurement, distribution, and returns management strategies. In an OC scenario, the various return channels available to the customers and their ability to switch between the channels complicates the decision-making (Bijmolt et al., 2021). This mandates the need for a closed-loop perspective in an OCSC, considering the evolving needs of the customers and the retailers. Also, returns have traditionally been considered an afterthought or a necessary evil. However, a well-designed reverse logistics system can not only aid in improved fulfillment and customer satisfaction but also provide a competitive edge to the retailer (Muir et al., 2019). Also, optimizing the forward and the reverse flows independently might not lead to an efficient supply chain. Hence, it is necessary to consider the forward and the reverse flows in conjunction to manage the incoming returns efficiently, optimize the forward flow, and integrate the returns into the forward flow at a faster rate (Sun & Tian, 2018).

The flow diagram in Fig. 1 considers the OCSC to be constituted of following three phases wherein each phase considers returns: (1) OC Purchase phase considering Returns (2) OC Delivery phase considering Returns (3) OC Returns Management phase. While there has been a systematic review on returns in OC in the past (De Borba et al., 2020), the authors lack the closed-loop perspective. Therefore, this systematic review adds to the literature on the subject area and expands the knowledge base from a novel perspective. Also, none of the other systematic reviews in the past aim at jointly addressing the omnichannel from a closed-loop perspective. The review aims at associating the themes identified from the systematic review with the phases in the flow diagram. This will help identify the research gaps in various phases. We categorize the identified themes in the review into the above-mentioned three phases.

2.1 OC Purchase Phase Considering Returns

This phase entails customers obtaining information and adopting a purchasing strategy, such as showrooming, webrooming, direct online purchase, and direct offline purchase. Showrooming involves customers visiting the physical store or the retailer's showroom to perceive the product information prior to making an online purchase (Bell et al., 2018; Gao & Su, 2017b). Webrooming refers to customers gathering information online before making a purchase offline (Gerea et al., 2021; Orús et al., 2019). On the retailer's end, the processes involve all those activities that aid the customer in the respective decisions. It encompasses deciding whether to open a showroom for information provision only or consider opening a physical store to also facilitate faster product availability in the offline channel, implementing various OC strategies such as BOPIS (Buy Online Pick-up In Store), BORIS (Buy Online Return in Store), STS (Ship to Store), C&C (Click and Collect). The selection of a suitable OC strategy aids in steering customers to the right channel, thus leading to efficient returns management through the reduction and avoidance of returns while also expanding cross-selling opportunities (Wollenburg et al., 2019). During the purchase phase, the retailer may use a variety of customer steering strategies to direct the consumer to the lowest cost-to-serve channel (Wollenburg et al., 2019, Wollenburg, Hübner et al., 2018) by using showrooming, webrooming, and various incentives to lower forward distribution costs and returns. The retailer also decides on the pricing strategies and return policies that further offer customer steering opportunities (Wollenburg, Holzapfel et al., 2018; Wollenburg, Hübner et al., 2018). Customer steering is of paramount concern since it has significant implications on various operational decisions such as warehouse management, logistics and inventory management, and other fulfillment decisions (Wollenburg, Holzapfel et al., 2018).

Fig. 1 Customer purchase and return journey in OC retailing

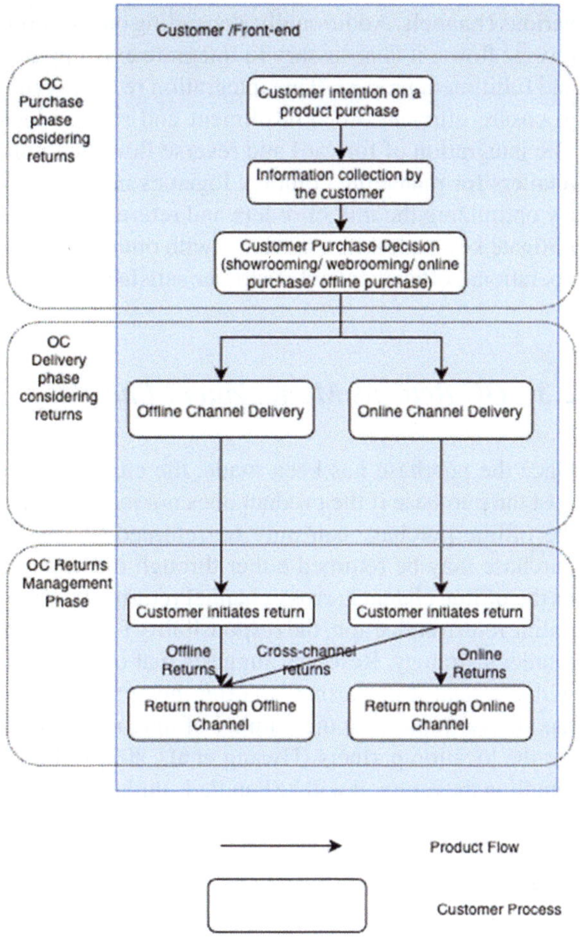

2.2 *OC Delivery Phase Considering Returns*

The customer completes an online purchase at the online store or an offline purchase at the physical store during the purchase phase. The purchase can be initiated in one channel while fulfilled through the other. For e.g., in the Buy Online Pickup In Store (BOPIS) omnichannel strategy, the customer places an order online but picks up the product in the store (Gao & Su, 2017a). So, while the order was initiated through the online channel, it was fulfilled through the offline channel. Such channel interaction demands a high level of integration between the channels and involves flexible leveraging of assets such as personnel, inventory, warehousing, and more (Taylor et al., 2019). The retailer, therefore, needs to consider various integrated fulfillment decisions such as inventory, warehouse, and logistics management for an OCSC, depending upon the OC strategy adopted as well as the returns from various channels. Additionally, depending on the adopted strategy for the forward and reverse flows, it is necessary to integrate existing infrastructure into the distribution and fulfillment process. This integration requires redefining the distribution network to ensure efficient order fulfillment and reverse logistics. (Melacini et al., 2018). The integration of forward and reverse flows in omnichannel retailing is crucial for retailers for managing complex logistics networks (Guerrero-Lorente et al., 2017). By optimizing the mix of orders and returns flows, retailers can minimize costs and mitigate key trade-offs associated with omnichannel models, thus enhancing overall operational efficiency and customer satisfaction.

2.3 OC Returns Management Phase

Once the purchase has been made, the customer can choose to return the product post the purchase if the product does not match the customer's needs. In most cases, the offline purchase can only be returned via the offline channel, but the online purchase may be returned either through the online channel (same channel return) or the offline channel (cross-channel return) (L. Yang et al., 2023). In the context of online return collection, the responsibility for collecting returns can vary based on the retailer's strategy. Research suggests that online retailers have started collaborating with physical store partners to offer the option of returning products in-store, which has shown to increase the number of unique customers, items sold, and net revenue for the location partners (Hwang et al., 2022). If a physical store is responsible for handling the return, it will be handled similarly to an offline return, with the exception that it will be picked up by a store representative. If the retailer takes responsibility for return collection, the retailer may employ in-house transportation services or outsource the return collection to a third-party logistics service provider. The return, in either scenario, will be transferred back to the Distribution Center. The retailer must optimize each of these decisions, which complicates the returns management processes in an OC scenario.

Post the return of the product by the customer, the physical store must determine whether to retain partial or complete inventory of the non-defective returned goods at the store to optimize resource utilization and rofitability (Wu & Lu, 2023). Returns that are defective will be transported back to the DC from the physical store. It is also critical to figure out the optimal quantities of non-defective returns to be returned to the DC and the optimal transshipment decisions between the stores. Non-defective returns are restocked by the Distribution Center. For the defective returns, decisions regarding repair, refurbishing, recycling, and selling in the secondary market need to be made. When selling in secondary markets, pricing strategies must be determined by taking into account the availability of identical items in both the primary and secondary markets.

3 Methodology/SLR Protocol

A fundamental aspect of academic research is to build on existing studies and knowledge to generate new knowledge (Snyder, 2019). The purpose of a literature review is to synthesize the relevant literature in order to analyze the knowledge base in a particular research field and to discover popular research streams and prospective research opportunities (Hochrein & Glock, 2012). Systematic literature reviews adopt literature search and selection procedures that are more reliable, rigorous, robust, and clearly defined, hence allowing for the repeatability of the sample and its assessment (Tranfield et al., 2003). Since the objectives of a Systematic Literature Review (SLR) perfectly meet our objectives, the systematic review approach was adopted.

Using the guidelines established by (Tranfield et al., 2003) and (Durach et al., 2017) for a systematic literature review, we undertake a systematic search for literature that addressed returns management and reverse logistics in OCSC. We further adopt the Prisma checklist proposed by (Liberati et al., 2009) to ensure transparency and accountability of the systematic review.

Due to the high degree of overlap across numerous databases, two of the most extensively used electronic databases, Web of Science and Scopus, were chosen for the study in order to reduce the likelihood of missing any pertinent information. To further emphasize high-quality research, we restrict our literature search to only ABDC-listed publications in the Scopus and Web of Science databases. An initial search was conducted to have a better grasp of the terminology utilised in the relevant literature. In the preliminary keyword search, the following terms were utilized: ((“omni-channel” OR “ omnichannel” OR “ omni channel”) AND (“returns” OR “ returns management” OR “ reverse logistics”)). From the preliminary search results, articles that addressed the subject of interest were recognized and extensively reviewed by all the authors in order to determine all the keywords required to search for the relevant literature. During the comprehensive review of the articles in the preliminary assessment, the authors noted the relevant keywords and added them to the search string. The final search strings shown in Table 1 were input into the Web of Science and Scopus databases to search for relevant literature. For both databases, search keywords were limited to “Title,” “Abstract,” and “Keywords.”

Inclusion Criteria:

Articles written in English language only are considered for review. Only journal articles were considered and conference proceedings, book chapters and all other forms of grey literature were excluded from the search. Literature available in electronic format (available in Scopus and Web of Science databases) was only considered. Papers addressing returns management or reverse logistics in OCSC particularly were considered relevant and included in the sample. We further restrict the review to only those articles published in ABDC indexed journals in the Scopus and Web of Science databases. Considering the focus on operational aspects of omnichannel supply chain, any articles indicating an obvious inclination to the marketing domain

Table1 Search terms used in Scopus and Web of Science databases

Database	Field of search	Search term	No of papers
Scopus	Title, Abstract, Keywords	TITLE-ABS-KEY ((("omni-channel" OR "omni channel" OR "omnichannel" OR "BOPIS" OR "BOPUS" OR "Buy online pickup in store" OR "Buy online pickup in physical store" OR "BORIS" OR "Buy Online Return in Store" OR "Buy Online Ship to Store" OR "Ship to Store" OR "Click and Collect" OR "Buy Online Ship from Store" OR "Ship from Store" OR "Cross-channel" OR "O2O" OR "Online to Offline" OR "Ship to locker") AND ("Return" OR "reverse" OR "Reverse Logistics" OR "cancel")) OR (("instore" OR "in store" OR "in-store") AND ("return" OR "returns")))	358
Web of Science	TS (Title, abstract, keywords)	TS = ((("omni-channel" OR "omni channel" OR "omnichannel" OR "BOPIS" OR "BOPUS" OR "Buy online pickup in store" OR "Buy online pickup in physical store" OR "BORIS" OR "Buy Online Return in Store" OR "Buy Online Ship to Store" OR "Ship to Store" OR "Click and Collect" OR "Buy Online Ship from Store" OR "Ship from Store" OR "Cross-channel" OR "O2O" OR "Online to Offline" OR "Ship to locker") AND ("Return" OR "reverse" OR "Reverse Logistics" OR cancel)) OR (("instore" OR "in store" OR "in-store") AND ("return" OR "returns")))	165

were excluded from the sample. The selection of articles in the sample was determined following a comprehensive and independent evaluation by each author. The inclusion of every article was individually considered by the research team members for discussion and eventually determined. Figure 1 indicates the flow chart that elucidates the process of obtaining the initial sample that was considered for the review. Table 2 indicates the exhaustive list of articles selected in the sample (Fig. 2).

4 Results & Discussions

General Overview of the Papers:

The 22 papers considered in the review are published between 2016 and 2023. This highlights the infancy stage of the research on returns management and reverse logistics in OCSC. The fact that only 22 articles were published in a period of 7 years is evidence of the lack of research in the subject area.

Table 2 List of articles included within the table and OC strategies and aspects addressed

			OC Strategies								Return/Cancellation		Specific aspect addressed
Phase	Theme	Paper	BOPIS	BORIS	Showrooming or ESBO or TSBO	SFS	STS	ROPS	Click and Collect	None	Return	Cancellation	
OC Purchase Phase considering returns	Pricing and Return Strategies	(D. Jin et al., 2020)		✓							✓		Strategic decision of BORP adoption and pricing strategies from a competitive perspective
		(Nageswaran et al., 2020)		✓							✓		Pricing and return policy decisions considering product valuation uncertainty
		(J. Liu & Xu, 2020)			✓						✓		Pricing and return policies considering showrooming

(continued)

Table 2 (continued)

Phase	Theme	Paper	OC Strategies								Return/Cancellation		Specific aspect addressed
			BOPIS	BORIS	Showrooming or ESBO or TSBO	SFS	STS	ROPS	Click and Collect	None	Return	Cancellation	
		(Mandal et al., 2021)		✓	✓						✓		Pricing and return penalties when adopting ESBO, BORS OC strategies
		(Jena & Meena, 2022)			✓						✓		Pricing and return strategies for price competing manufacturers adopting TSBO strategy

(continued)

Table 2 (continued)

			OC Strategies								Return/Cancellation		Specific aspect addressed
Phase	Theme	Paper	BOPIS	BORIS	Showrooming or ESBO or TSBO	SFS	STS	ROPS	Click and Collect	None	Return	Cancellation	
		(D. Jin & Huang, 2021)		✓							✓		Adoption of BORIS strategy by competing e-tailers and their pricing strategies
	Pricing and Inventory Strategies	(Zhang et al., 2018)						✓			✓	✓	Pricing and inventory decisions in ROPS considering order cancellations and product returns

(continued)

Table 2 (continued)

			OC Strategies								Return/Cancellation		Specific aspect addressed
Phase	Theme	Paper	BOPIS	BORIS	Showrooming or ESBO or TSBO	SFS	STS	ROPS	Click and Collect	None	Return	Cancellation	
		(Z. Li et al., 2021)		✓									Pricing and Inventory Strategies in BORIS
		(M. Jin et al., 2018)	✓					✓				✓	Pricing and radius of recommended service area decisions in BOPS scenario considering order cancellations

(continued)

Table 2 (continued)

			OC Strategies								Return/Cancellation		Specific aspect addressed
Phase	Theme	Paper	BOPIS	BORIS	Showrooming or ESBO or TSBO	SFS	STS	ROPS	Click and Collect	None	Return	Cancellation	
		(Y. Li, 2021)	✓								✓	✓	Pricing and Inventory Strategies in BOPS based on the influence of consumers' reference price effects considering order cancellations and product returns

(continued)

Table 2 (continued)

			OC Strategies								Return/Cancellation		Specific aspect addressed
Phase	Theme	Paper	BOPIS	BORIS	Showrooming or ESBO or TSBO	SFS	STS	ROPS	Click and Collect	None	Return	Cancellation	
		(Shi et al., 2018)	✓								✓		Pricing and ordering decisions for BOPS with pre-orders and returns
		(He et al., 2020)		✓							✓		Pricing and ordering decisions in BORIS

(continued)

Table 2 (continued)

			OC Strategies								Return/Cancellation		Specific aspect addressed
Phase	Theme	Paper	BOPIS	BORIS	Showrooming or ESBO or TSBO	SFS	STS	ROPS	Click and Collect	None	Return	Cancellation	
		(J. Liu & Xu, 2020)	✓								✓		Pricing and ordering decisions in BOPS considering returns
OC Delivery Considering Returns	Physical Network Design	(Gao et al., 2022)	✓	✓	✓						✓		Decision on the number of physical stores and their sizes considering the cross-channel returns in OC retailing

(continued)

Table 2 (continued)

Phase	Theme	Paper	OC Strategies								Return/Cancellation		Specific aspect addressed
			BOPIS	BORIS	Showrooming or ESBO or TSBO	SFS	STS	ROPS	Click and Collect	None	Return	Cancellation	
		(Bernon et al., 2016)		✓					✓		✓		Impact of OC retailing on returns management and the challenges and innovations associated with physical network and processes for OC returns

(continued)

Table 2 (continued)

			OC Strategies								Return/Cancellation		Specific aspect addressed
Phase	Theme	Paper	BOPIS	BORIS	Showrooming or ESBO or TSBO	SFS	STS	ROPS	Click and Collect	None	Return	Cancellation	
		(Guerrero-Lorente et al., 2020)								✓	✓		Last mile network design for a parcel carrier managing the online orders and pickup of returns for OC retailers

(continued)

Table 2 (continued)

Phase	Theme	Paper	OC Strategies								Return/Cancellation		Specific aspect addressed
			BOPIS	BORIS	Showrooming or ESBO or TSBO	SFS	STS	ROPS	Click and Collect	None	Return	Cancellation	
	Inventory Management	(G. Yang & Ji, 2022)		✓	✓						✓		Impact of cross-selling and optimal inventory management decisions in 3 returns management strategies

(continued)

Table 2 (continued)

			OC Strategies								Return/Cancellation		Specific aspect addressed
Phase	Theme	Paper	BOPIS	BORIS	Showrooming or ESBO or TSBO	SFS	STS	ROPS	Click and Collect	None	Return	Cancellation	
		(Goedhart et al., 2023)		✓							✓		Replenishment and rationing decisions for a physical store in an omnichannel setting considering multi-period sales dependent returns

(continued)

Table 2 (continued)

			OC Strategies								Return/Cancellation		Specific aspect addressed
Phase	Theme	Paper	BOPIS	BORIS	Showrooming or ESBO or TSBO	SFS	STS	ROPS	Click and Collect	None	Return	Cancellation	
		(L. Yang et al., 2023)		✓							✓		Impact of implementing BORIS strategy on retailer's store operations in terms of inventory management, customer base and profitability

(continued)

Table 2 (continued)

			OC Strategies								Return/Cancellation		Specific aspect addressed
Phase	Theme	Paper	BOPIS	BORIS	Showrooming or ESBO or TSBO	SFS	STS	ROPS	Click and Collect	None	Return	Cancellation	
OC Returns Management	Returns Processing / Returns Management Process	(Hübner et al., 2016)		✓							✓		Forward and backward concepts in distribution systems of OC retailing

(continued)

Table 2 (continued)

			OC Strategies								Return/Cancellation		Specific aspect addressed
Phase	Theme	Paper	BOPIS	BORIS	Showrooming or ESBO or TSBO	SFS	STS	ROPS	Click and Collect	None	Return	Cancellation	
		(Serkan Akturk et al., 2018)					✓				✓		Impact of introducing ship-to-store service on retailer sales and customer returns
		(Bijmolt et al., 2021)								✓	✓		Exploring customer steering opportunities for returns (backward flow) and returns prevention and processing strategies in OC retailing

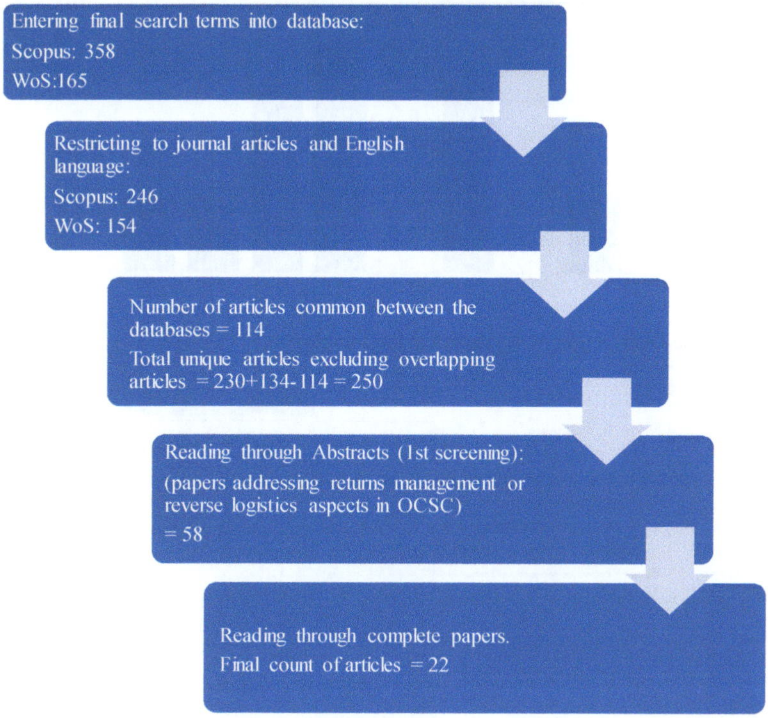

Fig. 2 Flowchart for selection of sample of papers

4.1 Descriptive Analysis

This section provides a descriptive analysis of the papers included in the sample for the systematic review. Figure 3 indicates the papers published per year within the sample. The figure indicates an increase in the trend of publication in the domain of returns management and reverse logistics in OCSC. While no papers were published in the years 2017 and 2019, there is an overall increasing trend observed in the number of publications each year until 2020. However, the number of publications have decreased over the years 2022 and 2023. Figure 4 indicates the graph for number of papers published in each journal. No consistent pattern was identified, as most journals published only one paper each. However, the European Journal of Operational Research published three articles, and the International Journal of Production Economics, Management Science, and Omega each published two articles. In order to identify the top 5 papers in the subject area, a simple citation score was determined from each of the paper by dividing the number of citations from google scholar with the number of years since the paper was published. The top 5 papers thus include (Bijmolt et al., 2021), (Hübner et al., 2016), (Nageswaran et al., 2020), (He et al., 2020) and (Bernon et al., 2016). Although in terms of pure number of citations, the top 5 papers include (Hübner et al., 2016), (Bernon et al., 2016), (M. Jin et al., 2018),

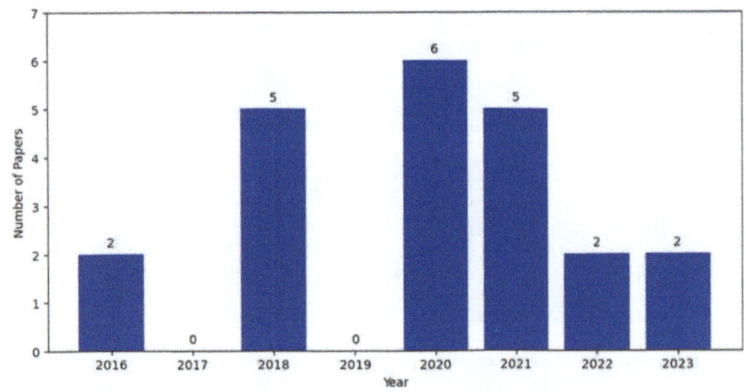

Fig. 3 Year-wise Distribution of the papers in the sample

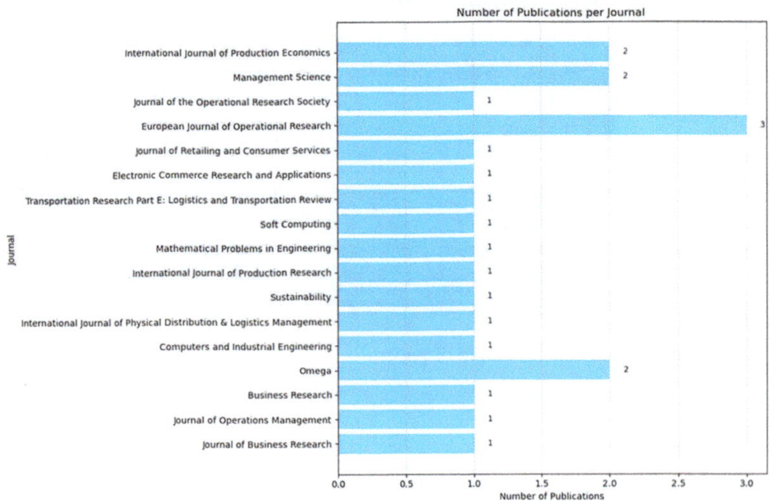

Fig. 4 Journal-wise Distribution of the papers in the sample

(Serkan Akturk et al., 2018) and (Bijmolt et al., 2021). Table 1 provides an overview of the various OC strategies studied in the articles considered in the sample.

4.2 *Categorization of Literature into Themes*

In this section, we categorize the literature on returns management and reverse logistics in OC into various themes depending on the nature of the decision problem addressed in the article. Each of the themes is further assigned to a phase of the customer product purchase and return journey in OC retailing, shown in Fig. 1. The several aspects investigated by the publications were identified after the papers were grouped thematically for a fuller comprehension of the studies' intricacies. This

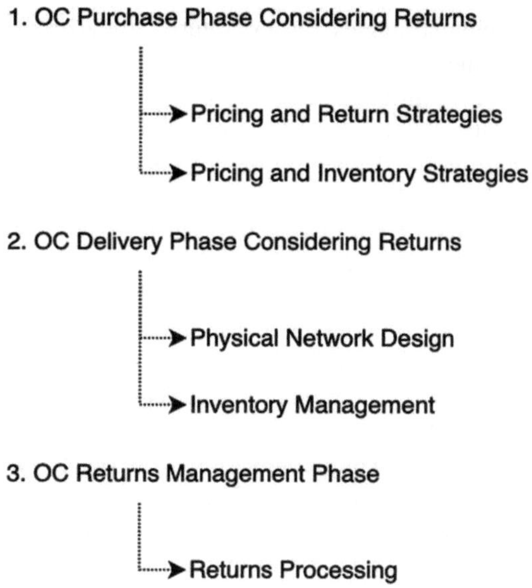

Fig. 5 Structure of themes based on phases in conceptual framework

section elucidates the scope and the nature of the decision problem and an in-depth analysis of the literature within the theme, thus enhancing the comprehension of the current state of the literature in the theme. This section answers RQ1 and RQ2. Figure 5 below indicates the various themes within each phase of the framework.

4.2.1 OC Purchase Phase Considering Returns

Pricing and Return Strategies

Pricing is one of the most significant factors that affect the customers' channel choice and, besides affecting the profit margins, also impacts the cannibalization intensity of showrooming since customers are driven to showrooming through the channel price gap (J. Liu & Xu, 2020). Besides, most customers (about 66%) review the return policy before making a purchase, and most retailers charge a restocking fee or return penalty in the range of 10% to 25% of the price of the product for an online return which discourages online purchase (Abdulla et al., 2019). Thus, customers are inevitably concerned with the pricing as well as the return policies as they are substitutable, which creates an obvious interlink between the decision problems (J. Liu & Xu, 2020; Mandal et al., 2021). Also, pricing and return policy decisions present customer steering opportunities and have implications for operational decisions such as warehousing, inventory, distribution, and returns management (Wollenburg et al., 2019; Wollenburg, Hübner et al., 2018). When customers decide from which retailer and from which channel to purchase, pricing strategies and the return policy offered by the retailers play a prominent role. Thus, retailers need to critically decide the pricing strategies and the return policy strategies in conjunction with the available channels.

In such a case, (D. Jin et al., 2020) determine the effect of BORIS adoption on the profit and on the pricing and online return fees of the retailer and identify equilibrium decisions made by retailers as well as the specific conditions under which retailers select a certain BORIS strategy configuration in a competitive scenario. In a competitive scenario, in order for one or both the retailers to offer BORIS as an equilibrium strategy, the products offered need to be significantly differentiated and the offline channel must be capable of salvaging returned products more efficiently than the online channel (D. Jin et al., 2020). However when retailers offer highly substitutable products or have online and offline channels that are equally efficient in salvaging returned products, they opt not to offer BORIS in an equilibrium (D. Jin et al., 2020).

Successful implementation of the BORIS strategy with the appropriate pricing and returns policies can significantly reduce the returns rate and the returns management cost while elevating customer satisfaction. The increasing competition among the retailers has also led to increasing competition in terms of return fees charged. A lenient return policy enhances customer perception of the services offered and reduce the perceived risk (Bijmolt et al., 2021), it, on the other hand, increases the returns rate and the associated complexity of managing the returns to a large extent (Bernon et al., 2016; Muir et al., 2019; Zhang et al., 2017). A strict return policy on the other hand can reduce the return rate but can create a reluctance to purchase among the customers and reduce the demand (Altug & Aydinliyim, 2016; Ofek et al., 2011). To address this issue, (Nageswaran et al., 2020) justify the generous full refund policies or partial refund policies adopted by various firms and state the conditions under which omnichannel retailers must offer a full refund or charge return fees for online returns. According to (Nageswaran et al., 2020), OC retailers that have competent salvage partners for the online returned products and with significant offline customer base should offer full refunds. Pricing and return strategies also present customer steering opportunities (Wollenburg et al., 2019; Wollenburg, Hübner et al., 2018). Retailers are incentivized to offer a full refund for products that customers can inspect in-store and that retailers with a large store network and appropriate instore returns salvaging opportunities would be better off charging a return fee to steer customers towards instore returns (Nageswaran et al., 2020).

Although a full refund can be generous and attractive to the customers, from an operational perspective, lenient return policies can lead to an absurdly high return rate. Thus it becomes imperative to offer optimal refunds to the customers. Certain retailers encourage customers for the showrooming strategy wherein the customers experience the product in the store but purchase online thus availing the advantages of both channels, i.e. experiencing the product in the offline channel and low cost order fulfilment through the online channel (Jing, 2018; J. Liu & Xu, 2020). Adoption of showrooming can boost profits and reduce product returns (J. Liu & Xu, 2020) since showrooms communicate the tactile information necessary for a customer to purchase the product without any effect on the fulfillment options in the showroom, thus simplifying the operations compared to a physical store (Bell et al., 2018). The channel price difference is the significant aspect that drives showrooming, whereby the retailer can maintain a price gap between the channels, thus promoting the shift in demand from offline to online channels and reducing the cost of fulfillment and returns (J. Liu & Xu, 2020). A differential pricing scheme coupled with well-designed return policies can benefit the retailers by adopting showrooming if the online channel has a fulfillment cost advantage and it can aid in the shift of demand from the offline channel (high-cost) to online channel (low-cost).

However, distinct products with varying attributes demand varying physical presence needs of the retailer. (Mandal et al., 2021) thus recommend optimal OC strategy of either establishing a showroom or a brick-and-mortar store based on product standardization and product valuation, its effect on the retailer's profit, and the optimal price and return penalties in the respective scenarios. (Mandal et al., 2021)'s findings suggest that for products with high return rates, the best strategy is to adopt showrooms and for high valuation and standardized products such as electrical appliances, retailers should adopt brick and mortar stores. If the reverse logistics cost is low, authors suggest that the retailer is better off selling through the online channel and in case of moderate reverse logistics costs, the optimal decision is to establish a showroom. (Jena & Meena, 2022) analyse a distinct showrooming scenario wherein a retailer sells products of two manufacturers online, however, displays products of only one manufacturer in the physical showroom, who bears the display cost while a return policy is adopted for the other manufacturer. The authors analyze the impact of such a TSBO (Test In Store Buy Online) strategy in a competitive scenario with two manufacturers and a single retailer on the pricing and total SC profit.

Pricing and Inventory Strategies

Several authors highlight the significance of the integrated inventory management for the available channels given the complex nature of the omnichannel supply chain (Kembro et al., 2018; Taylor et al., 2019). Due to the uncertainty in market demand, inventory management is a complex decision, but adopting OC strategies further complicates the decision-making process due to the interaction between the channels and the management of integrated inventory for the channels (Y. Li, 2021). Additionally, the communication of the inventory information and recommendation substitution products to the customers can assist in customer steering (Wollenburg et al., 2019; Wollenburg, Holzapfel et al., 2018). While pricing has a significant impact on customer channel choice (J. Liu & Xu, 2020), the pricing and inventory decisions when analysed together can present significant customer steering opportunities. Thus the optimal pricing and inventory decisions are crucial from the operational perspective since customer steering opportunities impact the operational decisions based on the channel to which the customers are steered (Wollenburg et al., 2019).

(Y. Li, 2021) examine how an OC retailer adopting BOPIS OC strategy decides on the pricing and inventory decisions while taking reference price effects on the customers into account. Customers can either cancel an order before the payment is made or return the product after payment if the product does not fit the customer needs. According to authors, when consumers are more sensitive to the difference between the selling and reference prices or if customers have a lengthy memory of prior pricing, the equilibrium selling price and ending inventory levels at the end of the period fall. (Shi et al., 2018) address the optimal pricing and ordering strategies for BOPIS strategy with pre-orders while considering returns and state that retailers shouldnt always adopt BOPIS with pre-orders. The profit of the retailer through the BOPIS strategy for preorders as compared to pure physical store selling when the unit production cost is small. The BOPIS strategy with preorders is beneficial if the degree of demand uncertainty exceeds the threshold. When a returned product can be resold (or cannot be resold), the unit production cost (salvage cost) and the refund decide whether the retailer should allow returns in a BOPIS scenario with pre-orders.

In the BOPIS implementation, customers are sensitive to the distance from the store for the order pickup. Customers are reluctant to utilize the BOPIS service if the distance from the store is long (M. Jin et al., 2018). In order to address this issue, (M. Jin et al., 2018) explore the pricing and the radius of recommended service area to fulfill the online and offline orders in a BOPIS scenario considering order cancellation. In the case of BOPIS, order cancellations are treated as returns and processed similarly (Bernon et al., 2016). They also compare BOPIS and ROPS OC strategies considering various order cancellation policies offered by the retailer and find that moderate cancellation policy is the most profitable in a BOPIS scenario and the profit of the retailer is equal in the BOPIS and the ROPS scenarios with moderate cancellation policy.

(J. Liu & Xu, 2020) construct expected profit models before and after BOPIS is offered by the retailer considering consumer surplus and purchase ratio in different channels and determine the optimal pricing and ordering decisions. The authors then conduct a comparative analysis of the pricing and ordering decisions before and after the BOPIS adoption and suggest that retailers should implement BOPIS since it assists in addressing diverse needs of the customers, boosts cross-selling, demand and customer loyalty, and promotes the sustainable development. However, (Zhang et al., 2018) argue that omnichannel strategies are not always promising for the retailers and identify the various conditions under which omnichannel strategies are beneficial to retailers. (Zhang et al., 2018) evaluate whether an online retailer should open physical stores for OC adoption by evaluating the value of opening a store for the online retailer. They investigate the pricing and inventory decisions in the Reserve Online Pickup in Store (ROPS) strategy to address the trade-off between high number of cancellations and benefits to customers with ROPS and find that the prices are lower in the omnichannel strategy than the online only single channel strategy and the omnichannel strategy minimizes the returns while increasing the cancellations.

(Y. Li, 2021) study the pricing and inventory strategies for a retailer that adopts BORIS strategy and its effect on operational strategies of the firm considering consumers' and firms' rational expectation behavior and random demand. The adoption of BORIS is advantageous to the retailer when both instore and online returns costs are low (Y. Li, 2021). The inflow of returns to the store in a BORIS scenario can significantly affect the pricing and inventory decisions for the store given the uncertainty of returns along with uncertainty in demand. It becomes necessary to address the disposition of the returned products wherein some products are reused through their sale in the secondary markets as refurbished products. The store needs to determine the pricing strategies considering the sale of similar products in the primary and the secondary markets and the inventory decisions considering the inventory of returned products. (He et al., 2020) thus developed a news-vendor model to determine the optimal pricing and ordering decisions for a retailer selling through online and offline channels while considering the resale of products returned by the customers as refurbished products. Their findings indicate no change in the optimal price of the new product while a decrease in the price of the refurbished product under certain conditions, and an increase in the order quantity of the online channel in the omnichannel scenario than the online-only single channel scenario while the order quantity of the store is reduced.

4.2.2 OC Delivery Phase Considering Returns

Physical Network Design

Effective design of the reverse logistics network can lead to significant savings in costs through optimization of the transportation, inventory, order processing, and warehousing (Amini et al., 2005). (Bernon et al., 2016) highlight two management aspects of network design for returns in OC: Customer accessibility to return entry points and poor integration of the return channels due to an increase in complexity. The network design, while reducing the costs for the retailer, must also ensure that the physical stores are accessible to the customers for the purchase as well as the return transactions. In the case of omni-channel, the retailer must also ensure integration of the available channels for a seamless customer experience as well as a responsive and cost-efficient supply chain. Owing to the decisions that depend on the physical network design and the new challenges in the OC context, it becomes a problem of paramount concern for omnichannel retailers.

Retailers with a larger store network can leverage their own stores for forward fulfillment and reverse logistics, while retailers with limited store coverage can partner with third-party service providers (Bernon et al., 2016). (Bernon et al., 2016) in their study find that even retailers with large store networks leveraged third-party services to offer their customer-wide range of choice for the returns. Thus, the network design problem also encompasses the decision of whether to outsource and what processes to outsource and to what extent. As a consequence of integration in an OCSC, the network design problem needs to be addressed for the forward and reverse flows in conjunction for an optimal network design. Despite the emerging studies on omnichannel retailing there are no empirical or analytical studies to the best of authors knowledge that study the network design in an OC context.

While network design comes under the purview of strategic decisions, last-mile delivery is an operational decision. Last mile delivery incurs tremendous cost owing to the complexities in the delivery and returns. Additionally, last mile network and distribution plays a pivotal role in the efficient forward and reverse distribution in an OCSC. (Guerrero-Lorente et al., 2020) analyze last mile network design for a parcel carrier managing the online orders from OC retailers. They develop a mixed integer program (MIP) for the last mile network design problem and consider various types of facilities such as local distribution centers, intermediate depots, parcel offices, collection points, stores and kiosks and Automated Parcel Stations (APS). The authors however address the problem for a third-party service provider managing the online orders for an omnichannel retailer. The last-mile delivery and returns planning for omnichannel retailers becomes quite complex due to the customers ability to seamlessly switch between the channels. If a customer has placed an order online, he can either receive the product at his desired location or pick up the product at the store and can raise a return request either to be picked up at his desired location or returned to the store. Considering such complexities, last mile delivery and returns planning require meticulous planning and execution to reduce the costs while offering customers flexibility. The literature, however, lacks the addressal of last mile network planning.

Omnichannel also has redefined the role of stores in the retailing. Traditionally stores were utilised for pure offline purchases and returns. However, with omnichannnl retailing stores can act as showrooms for products to be displayed (but not for sale), or for the fulfilment of online orders through the store for strategies such as STS or BOPIS and for return of online order for strategies such as BORIS. The literature needs to study the emerging role of physical stores in OC retailing in the last mile network planning for delivery as well as returns (Hübner et al., 2022).

(Gao et al., 2022) address how omnichannel retailers must deal with the problem of identifying the number and size of physical stores. Findings suggest retailers with high return rate for online purchases are incentivized to have fewer but larger physical stores and with the increase in the convenience of online shopping, the retailer may prefer to adopt more physical stores but smaller in size. The authors state that the optimal decision for retailers in the case of showrooming strategy would be to reduce the number of physical stores and increase their size, increase the number in the case when the stores are used as fulfillment centers and decrease the store size when stores are used to collect returns. (Hübner et al., 2022) conduct a systematic review that elucidates the importance of stores in an OCSC. As online shopping is faster and convenient with the advancement in technology, physical stores are equally gaining importance and playing a more significant role contrary to the belief that physical stores will be extinct (Gao et al., 2022). More analytical studies are needed to understand how utilising stores as forward fulfilment centers and/or return collection centers can optimize the forward and the reverse flows in an OCSC while increasing convenience to the customers. Utilizing stores for forward fulfillment and returns collection is a complex task since it could create a stock out condition in one store while there could be mounting inventories in the other. To address this issue, inventory transshipment decisions between stores and between stores and distribution centers need to be studied.

Inventory Management

Inventory management in omnichannel retailing is one of the crucial challenges to the implementation of omnichannel retailing (Arslan et al., 2021). Additionally, integrating returns to inventory management in an omnichannel scenario is essential for optimizing operations and enhancing profitability. Research suggests that integrating forward and reverse logistics by adopting policies like Buy Online Return in Store (BORIS) can lead to improved logistical efficiency and profitability (L. Yang et al., 2023). The rise of omni-channel retailing has led to higher return rates for online sales, emphasizing the importance of mature returns management processes and inventory management to offer a seamless solution (Bernon et al., 2016).

(G. Yang & Ji, 2022) assess the impact of cross-selling on sales and profitability in the case of 3 returns management strategies: (a) BORS (b) Return Insurance (c) Virtual try-on Experience. Authors determine the optimal inventory in case of each of these 3 strategies considering cross-selling. To increase the profitability of the retailers, they suggest a combination of profits from cross-selling and profits from initial sale of products. The authors also highlight the importance and benefits of combining innovative strategies and cross-selling to effectively manage returns and increase the profitability of the retailer. (Goedhart et al., 2023) analyze the replenishment and rationing decisions for a physical store catering to the online and offline channels in an omnichannel setting, considering multi-period sales-dependent returns. For small scale instances, authors suggest value iteration solution, while for large scale instances, authors propose a deep reinforcement learning algorithm. (L. Yang et al., 2023) explore how implementing Buy Online Return in Store (BORIS) affects store operations. They focus on decisions regarding inventory management, effect on customer base and profit outcomes. Their findings indicate that BORIS can attract new customers, prompt existing customers to switch channels, and reduce in-store inventory levels. However, the suitability of BORIS varies by product category. The study also reveals that customers are more likely to use BORIS when return penalties and return rates are low.

4.2.3 OC Returns Management Phase

Returns Processing / Returns Management Process

Once a product is returned by the customer, the retailer has to ensure that the returned product reaches the desired destination rather than ending up in the landfill. If the product is non-defective, it can be merged into the saleable inventory and if it is defective, it will undergo a certain recovery process before it is resold in the secondary market or is recycled. While customers expect a faster refund post the return (Gawor & Hoberg, 2019), the retailer expects faster restocking of the product into the saleable inventory (Muir et al., 2019). In an OC scenario, the returns management is more complex than ever owing to the liberal return policies (Bernon et al., 2016), multiple return options (in-store returns, online channel return, Courier,

Express and Parcel (CEP) return) (Kembro et al., 2018), uncertainties associated with the return and the channel of return (Xu & Jackson, 2019).

(Hübner et al., 2016) provide a holistic view of the forward and backward flows. They identify the issues relevant to logistics services and costs in OC retailing and highlight the need for faster and efficient processing of returns from the customer and the retailer perspective since the customer expects a faster refund, while the retailer expects the faster integration of the returned product back into the forward flow.

(Hübner et al., 2016) provide a typology for returns processing in OC retailing considering various return modes such as CEP return and store return as well as various returns inventory handling strategies such as adding the returns to store inventory vs. shipping the returns to Distribution Centers (DCs)/ Return Centers (RCs). While returning a product through CEP simplifies the overall process, it incurs additional transportation costs and time to integrate the product back into the sellable inventory. On the other hand, the store return option provides better convenience to the customer and better access to feedback, potential for cross-selling and faster integration of the returned product into the sellable inventory for the retailer (Hübner et al., 2016). In the case of store returns it becomes mandatory to conduct a preliminary evaluation to determine the condition of the product since a non-defective return can be immediately added to the store inventory but a defective return would require a certain amount of reprocessing depending upon the defect in the product. This mandates the need to study the returns processing structure in a manner that will integrate the returned product back into the saleable inventory at a faster rate. (Hübner et al., 2016)s findings indicate that only half of the retailers in their study permitted in-store returns, and the reluctance of the retailers was owing to the extra labor at the stores, complications in the refund, and insufficient IT infrastructure. The retailer must decide whether to process the non-defective returns at the store or to forward the inventory to the DC for processing. Also, (Hübner et al., 2016) highlight the need for developing the complicated IT infrastructure necessary to optimize the OCSC.

Adoption of various OC strategies, such as BOPIS, STS, Click and Collect, etc., can have a significant impact on the returns in an OC. One such study was conducted by (Serkan Akturk et al., 2018). They assess the impact of introducing ship-to-store service on retailer sales and customer returns across multiple channels for an OC jewelry retailer and identify an increase in cross-channel returns due to the introduction of ship-to-store and also an increase in cross-selling at the physical store. Authors report an increase in returns for the high-value items at the physical store, however, observe no change in the returns for online purchases. The authors observed a shift of customers from the online channel to the offline channel for high-value purchases while for the low-value purchases, customers preferred to adopt the STS strategy. Such a study can provide valuable insights that assist in the returns management decisions. However, the study was conducted for a jewelry retailer which is a segment with products of high value and the customer purchase and return behavior may be different owing to the high product valuation uncertainty for this product category. Hence, the findings cannot be extrapolated to retailers dealing with other product categories, such as electronic, grocery, and apparel, with differing attributes and valuation uncertainties.

(Hübner et al., 2016) discuss the various returns processing locations such as the store for faster integration into saleable inventory or at the central DC or an independent Return Center (RC). (Hübner et al., 2016) state that the most common strategy to process returns involves processing the returns at the central DC with dedicated returns processing section. This is advantageous from the point of view that the returned products can be reprocessed and directly integrated into the online channel inventory, and flexible workforce scheduling can be adopted based on the volume of returns. The decision of returns processing location in conjunction with the forward flow planning is pivotal to ensure that the returns are integrated into the forward flow as quick as possible and no returns remain unprocessed. In-store returns are equally beneficial for the retailer as well as the customer since the customer receives an instant refund, and the retailer can add the returned product to the store inventory almost instantaneously (Bijmolt et al., 2021). When the amount of returns is overwhelming, retailers tend to adopt independent Returns Centers (RC), which are specialized returns processing entities often used by fashion retailers owing to their high return rates and the associated potential cost-saving opportunities (Hübner et al., 2016). Independent RCs lengthen lead times, posing further challenges due to the limited time demand for fashion products, which leads to a loss of sales. Thus, retailers need to address the trade-off between cost savings and faster processing in the returns processing locations decision. It is imperative to address the OC network design considering the forward and the reverse flows to optimize the forward fulfillment and the reverse logistics in conjunction.

(Bijmolt et al., 2021) highlight the customer steering opportunities for returns reverse flows in an OC setting. Steering customers to in-store returns can lead to significant cost savings due to its low cost-to-serve nature and also the cross-selling opportunities that arise when customers visit the store to return a product. When a customer returns a product purchased online to the offline physical store (i.e. cross-channel return), it reduces the transportation and handling costs incurred in reverse pickup of the product from the customer and transporting it back to the DC or the RC, the associated labor cost, ensures quicker integration of the returned product back into the saleable inventory and also provides additional cross-selling opportunities. When in the store, the customer can either opt to exchange the product for another or purchase other products while in the store. Certain customers prefer online returns over in-store returns, which mandates the need to study customer steering opportunities for the return channel in an OCSC. While there are studies that study customer steering opportunities in the forward channel (Wollenburg, Hübner et al., 2018), none of the studies explore the customer steering opportunities in the return channel for an OCSC.

The speed with which returns are processed is a major concern for the retailers as well as customers (Bijmolt et al., 2021; Hübner et al., 2016). The speed of processing returns affects the market value of the product returns in particular the seasonal products since the longer the time required for processing the return, the more the capital tied up (Bijmolt et al., 2021). While customer are more concerned about the refund, the retailers interest lies in the speed with which the returns can be added to the saleable inventory. The strategic decisions of whether to offer reprocessed returns at the physical or online stores is indispensable (Hübner et al., 2016) since any unsold returns contributes to the lost sales and mounting unsold returns and the returns processing locations significantly effects the speed. This decision affects the forward distribution of the primary market since similar products might be available at competitive prices in both, the primary and the secondary markets. The effect of offering reprocessed returned products alongside fresh products needs to be studied in an OC context. Some retailers prefer to offer reprocessed returns only in the physical store since online customers do not accept any signs of usage of the products, which can further increase the returns, while other retailers offer the reprocessed returns online at a reduced price. For retailers offering the reprocessed returns in the store, the products can be repurposed at the store thus saving transportation and handling costs (Hübner et al., 2016). Only 10–20% of the original value of returns can be recovered (Ofek et al., 2011). Retailers are putting in significant efforts to recoup as much value from a returned product as possible. However, given the complexities associated with returns management and the associated reprocessing of the returns for reselling, most returns end up in landfills (Duong et al., 2022). In order to avoid returns ending up in the landfill from a sustainability perspective, it is exigent that literature addresses this issue.

When a returned product is received by the retailer at the physical store or the DC or RC, the next step involves gatekeeping of the returns. Gatekeeping is vital to avoid re-selling of inferior quality products and fraudulent returns while addressing the trade-off between thoroughness and cost (Bijmolt et al., 2021). Post the inspection in the gatekeeping, the product can be either directly added to the saleable inventory, directly resold in the secondary market, directed for reprocessing operations such as repair, refurbishment, or remanufacture, and then sold in the secondary market, or disposed off completely (recycle) (Bijmolt et al., 2021). Apart from the decision of where to process the returns, the retailer needs to decide on the location of gatekeeping operations as well. Gatekeeping can be either centralised at the central DC/RC or decentralized at the store locations. Centralized gatekeeping can lead to savings in labor cost, increase the refund processing time, increase the time required to integrate the returned product into saleable inventory and increase the logistics costs, while decentralised gatekeeping can increase the labor cost, reduce the refund processing time, reduce the time required to integrate the returned product back into the saleable inventory and reduce the transportation cost (Muir et al., 2019). Thus, the retailer needs to address the trade-off between cost and service to decide on gatekeeping location. The literature lacks studies addressing this key decision in an OC context. Gatekeeping of returns is a deciding factor that decides the future processes of the product and how the product does not end up in the landfill.

5 Future Research Directions

The findings of the systematic review of the literature on returns management and reverse logistics in OCSC clearly indicate a significant scarcity of literature. There is an increasing need for studies in the subject area. However, the complexities associated with OC retailing demand studies differ significantly from the previous studies that addressed single-channel or dual/multi-channel supply chains (Hole et al., 2019; Raza & Govindaluri, 2021). Also, while past studies did address returns management and reverse logistics in OC, none of the studies highlighted the importance of addressing the forward and reverse flows in conjunction. We thus identify the need to adopt a closed-loop perspective in OCSC and propose a framework shown in Fig. 6. With this objective, we aim to propose a framework that considers returns during every phase of the customer journey. The aim of the framework is to optimize the forward and the reverse flows in conjunction for a more optimized OCSC as opposed to a standalone analysis of each of the flows. Since OC retailing provides customers access to all the available channels and the ability to switch between the channels seamlessly, it is imperative to consider the customer processes and the corresponding back-end or retailer processes simultaneously to understand the implications of the customer processes on retailer processes and vice versa. The conceptual framework intends to emphasize the fragmented nature of the literature on this subject and simultaneously provide future research directions to advance the knowledge systematically. The framework is proposed based on the Customer purchase and return journey in OC retailing flow diagram indicated in Fig. 1. This section answers RQ3.

The Framework consists of the same three phases indicated in the flow diagram in Fig. 1. However, it includes the retailer back-end processes that need to be undertaken for the corresponding customer process. We elaborate on each of the retailer back-end processes in Sect. 5.1. Table 3 summarizes the future research questions for each of the themes identified within the respective phases in the conceptual framework.

5.1 OC Purchase Phase Considering Returns

5.1.1 Channel Integration Through Technology Adoption

In order to provide seamless OC experience to the customers considering the complexities associated with channel integration in an OCSC, there is a need for advanced integrated IT infrastructure (Wollenburg, Holzapfel et al., 2018; Wollenburg, Hübner et al., 2018). (Wollenburg, Hübner et al., 2018) highlight the need of integrated infrastructure to facilitate cross-channel fulfillment and data transparency in an OCSC. Additionally, the need to address the OCSC from a closed-loop perspective further demands the adoption of improved IT infrastructure. The channel-integrated IT infrastructure must support the integration of warehousing and inventory and provide customers with the necessary cross-channel information

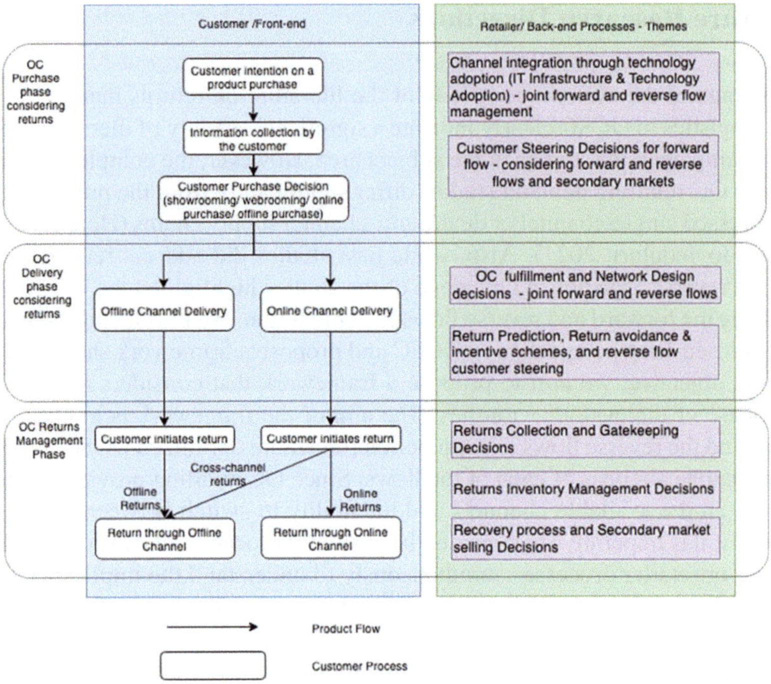

Fig. 6 Conceptual framework for future research directions

Table 3 Future research directions

Phase	Theme	Future research directions
OC Purchase Phase	Channel integration through technology adoption	2.1.1.1. What are the advanced IT infrastructure requirements to facilitate cross-channel fulfilment and data transparency in an Omnichannel Closed Loop Supply Chain (OCCLSC)? 2.1.1.2. How can the warehousing, inventory and logistics be integrated through IT infrastructure to provide customers with the necessary real-time cross-channel information for a seamless OC experience and to reduce returns? 2.1.1.3. How can technologies such as AR/VR be adopted in an OCCLSC to reduce the returns through mitigation of consumer fit uncertainty? 2.1.1.4. How can block-chain technology improve the traceability in an OCCLSC?

(continued)

Table 3 (continued)

Phase	Theme	Future research directions
	Customer steering decisions for forward flow - considering forward and reverse flows and secondary markets	2.1.1.4.1. What are the implications of various OC strategies such as Click and collect, Click and reserve, ship to store, ship form store, etc. on the pricing and return strategies? 2.1.1.4.2. What are the effects of order cancellations on pricing and return policies? 2.1.1.4.3. What are the pricing and inventory strategies in an OCCLSC considering the sale of refurbished products in the secondary market? 2.1.1.5. What are the effects of order cancellations on pricing and inventory strategies of different types of retailers (e.g. grocery, electronics, fashion retailers etc.)? 2.1.1.6. What are the pricing and ordering strategies considering the defective and non-defective returns received by a retailer? 2.1.1.7. What are the customer steering opportunities in the case of Ship from Store, Ship to Store, Reserve Online Pickup in Store and Click and Collect OC strategies? 2.1.1.8. What are the effects of order cancellations on the customer steering opportunities for various OC strategies? 2.1.1.9. What are the optimal returns policy decisions in an OC scenario considering the order of multiple versions of same product to retain the fit product and return the misfit ones? 2.1.1.10. What are the OC strategy adoption decisions, pricing strategies and return policy decisions considering multiple products and bundle of products? 2.1.1.11. What are the customer steering opportunities considering intra-product showrooming and inter-product showrooming and their implications on OC strategy adoption decisions, pricing strategies and return policy decisions? 2.1.1.12. What are the customer steering opportunities in a competitive scenario considering multiple players and multiple products? 2.1.1.13. What is the role of the manufacturer in the strategy adoption, pricing strategies and return policy decisions in an omnichannel scenario?

(continued)

Table 3 (continued)

Phase	Theme	Future research directions
OC delivery phase	OC fulfilment and Network design decisions – joint forward and reverse flows	2.1.1.13.4. What are the optimal inventory management, warehousing and transportation decisions for the hybrid OC physical store and the hybrid OC Distribution Centres considering the reverse flows for an efficient OCCLSC? 2.1.1.13.5. What are the optimal online delivery decisions (delivery by the nearest physical store or inhouse transportation services or outsourcing to 3rd party logistics) considering the reverse pickup and flow of returns? 2.1.1.13.6. What are the effects of implementing OC strategies such as BOPIS, STS, C&C, etc. on the fulfilment decisions of the entire supply chain? 2.1.1.13.7. What are the optimal last mile delivery decisions considering the delivery of products and return collections in an OCCLSC? 2.1.1.13.8. What is the network design for an OCCLSC? 2.1.1.13.9. What are the optimal location inventory decisions in an OCCLSC? 2.1.1.13.10. What is the effect of order cancellations on the inventory, logistics and warehousing strategies of various types of retailers? 2.1.1.13.11. What are the optimal fulfilment decisions considering the defective and non-defective returns? 2.1.1.13.12. Analysis of the joint implementation of various OC strategies 2.1.1.13.13. Analysis of joint implementation of various OC strategies considering the forward and the reverse flows in an OCCLSC 2.1.1.13.14. What are assortment and bundling strategies to be adopted in an OCCLSC?

(continued)

Table 3 (continued)

Phase	Theme	Future research directions
	Return prediction, return avoidance, incentive Schemes and reverse flow customer steering	1. Prediction of returns, channel of return, quality of return, category of product return and the probability of fraudulent return in an OC scenario 2. What are the incentive schemes to reduce/ avoid returns? 3. What are the incentive schemes to avoid/ reduce fraudulent returns? 4. What are the reverse flow customer steering opportunities in an OCCLSC? 5. What are the back-end requirements of reverse flow customer steering? 6. How can the implementation of various steering methods, the associated costs and actual benefit of implementing the strategies be quantified?
OC returns management phase	Return collection and gatekeeping decisions	1. What are the optimal gatekeeping decisions for returns in an OCCLSC? 2. Under what conditions should a retailer adopt centralized or decentralized gatekeeping of returns? 3. What are the optimal gatekeeping decisions (centralized vs. decentralized) for different types of retailers? 4. What are the optimal returns collection decisions (pickup by the nearest store or inhouse transportation or CEP carrier or outsourcing to 3PL service provider) considering the forward flow in an OCCLSC?

(continued)

Table 3 (continued)

Phase	Theme	Future research directions
	Returns inventory management decisions	1. What are the optimal reverse flow inventory management strategies in an OCCLSC? 2. What are the optimal inventory rebalancing strategies in an OCCLSC? 3. What are the store- to store and DC to DC inventory transshipment decisions in an OCCLSC? 4. What are the various innovative strategies to integrate the returns products back into the saleable inventory at a faster rate?
	Recovery process and secondary market selling decisions	1. What are optimal decisions on repairing, refurbishing, recycling and selling in secondary markets based on the condition of the returned product in an OCCLSC? 2. What are the optimal decisions on selling recovered returned products in secondary markets? 3. What are the innovative and optimal strategies to recover the defective returned products to ensure a Sustainable OCCLSC (SOCCLSC) and integrate the recovered product back into the saleable inventory at a faster rate? 4. Analysis of the social aspect of sustainability in a SOCCLSC

so as to offer a seamless OC experience (Hübner et al., 2016). Future studies should explore the needs of integrated IT infrastructure needed to manage an Omnichannel Closed Loop Supply Chain (OCCLSC). Future research should also explore the IT infrastructure needed to present the customers with real-time cross-channel inventory information so as to reduce returns (Bijmolt et al., 2021).

The adoption of various technologies in the Supply Chain domain is growing rapidly. Technologies such as Virtual Reality (VR), Augmented Reality (AR), Blockchain etc. possess tremendous potential to be adopted in Omnichannel Supply Chain (Cai & Lo, 2020). Technologies such as AR and VR assist in reducing customer fit uncertainty (Bijmolt et al., 2021). Future studies should explore how technologies such as AR and VR can reduce and avoid returns through the mitigation of consumer fit uncertainty by providing customers with a store-like experience (Bijmolt et al., 2021). (Bijmolt et al., 2021) highlight the need for modern technologies to trace the products in a Supply Chain. One such traceability system is blockchain technology. Blockchain technology can assist in the secured and reliable tracking of products in an OCSC. Given the complex situation due to the integrated channels and also the information of the closed-loop supply chain, blockchain technologies can significantly improve the traceability of the products in an OCCLSC. Furthermore, Internet of Things (IoT) devices facilitate real-time tracking of returned products and enhance the visibility and control over return processes, leading to efficient processing and restocking of returned products (Thaichon et al., 2024). Future research should explore the use of blockchain and IoT to enhance traceability in an OCCLSC.

5.1.2 Customer Steering Decisions for Forward Flow—Considering Forward and Reverse Flows and Secondary Markets

(Wollenburg et al., 2019) emphasize on the importance of customer steering during the information collection phase of the customer purchase journey. The customer steering decisions comprise of pricing strategies, OC strategy adoption and the return policy decisions (Wollenburg, Holzapfel et al., 2018). These aspects significantly influence the customers' channel choice during the purchase phase, and hence retailers can modify these strategies to steer the customers to a low-cost-to-serve channel (Wollenburg et al., 2019; Wollenburg, Holzapfel et al., 2018). The literature on pricing and return strategies in OC majorly focuses on only two strategies in OC retailing namely BORIS and Showrooming. Future studies can explore the pricing and return strategies for other OC strategies such as Click and collect, Click and reserve, Ship to Store, Ship form Store, etc. Order cancellations are usually considered returns and treated similarly (Bernon et al., 2016). Future research can analyse the effect of order cancellations on the pricing and return policies.

The literature on Pricing and Inventory Strategies explores order cancellation and considers BOPIS and ROPS strategies. It even so lacks the analysis of pricing and inventory strategies considering refurbished products and the effects of order cancellations on different types of retailers. Future studies can explore the effect of order cancellations on the pricing and inventory strategies of different types of retailers such as grocery, fashion, electronics and others. (He et al., 2020) explore the pricing and ordering decisions considering the sale of returned products as refurbished ones. However, in reality, the returns consist of defective and non-defective returns. The non-defective returns can be directly resold, while the defective returns need to undergo certain processing before they can be resold. Future studies can explore the pricing and return strategies considering the defective and the non-defective returns which reflect a more realistic scenario. Table 2 clearly highlights the lack of research on customer steering opportunities in the areas of Ship from Store (SFS) (He et al., 2020; M. Jin et al., 2018), Ship to Store (STS) (He et al., 2020; M. Jin et al., 2018), Reserve Online Pickup in Store (ROPS) and Click and Collect (C&C). Table 2 also highlights the lack of research on customer steering opportunities considering the order cancellations in ROPS, BOPIS, and Click and Collect and Ship-to-store strategies.

Another area for further investigation can be exploring the return policies in an OC scenario, considering customers ordering multiple versions of the same product to retain the fit product while returning the misfit ones (Nageswaran et al.,

2020). Most studies explore customer steering opportunities, assuming the sale of a single product. Exploring the adoption of OC strategies, pricing strategies, and return policies considering multiple products and bundles of products could be a valuable extension (Mandal et al., 2021). There is a lack of research exploring customer steering opportunities considering intra-product showrooming and inter-product showrooming and their implications on OC strategy adoption decisions, pricing strategies and return policy decisions. Intra-product showrooming implies a customer inspecting a product offline and buying the same one through the online channel, while inter-product showrooming implies a customer inspecting a product in the showroom and buying a different or similar product through the online channel (R. Li, 2020). Intra and inter-product showrooming are relatively unexplored from a returns management perspective in OCSC. Analysing the customer steering opportunities in a competitive scenario considering multiple players and multiple products could be an interesting future research direction (Jena & Meena, 2022). Additionally, exploring the role of the manufacturer in the OC strategy adoption, pricing strategies, and return policy decisions in an omnichannel scenario will be an exciting subject for future investigation (Mandal et al., 2021).

5.2 OC Delivery Phase Considering Returns

5.2.1 OC Fulfillment and Network Design Decisions—Joint Forward and Reverse Flows

The systematic review distinctly highlights the scarcity of literature identifying the forward fulfillment decisions considering the reverse flow of returned products. In order to adopt a closed-loop perspective for the OCSC, it is imperative to consider the reverse flows of returns during every phase of the OCSC. Future studies should explore the forward flow inventory management, warehousing and transportation decisions for the hybrid OC physical store and the hybrid OC Distribution Centers considering the reverse flows. Despite the serious implications of these decision problems and the complexities of integrated inventory management, warehousing, and transportation, these aspects remain unexplored in the literature to the best of the authors' knowledge. Also, most studies in the literature explore the addition of an online channel by a physical store retailer. Future works can explore the adoption of OC strategies by online retailers adding an offline channel (Gao et al., 2022). Decisions such as the delivery of online orders by the nearest physical store or the inhouse transportation services or outsourcing to a third party logistics provider considering the reverse pickup and flow of returns is a worthy topic for future research. The effect of various OC strategies such as BOPIS, STS, C&C, etc. on the fulfilment decisions of the entire supply chain is another area of future research (Mandal et al., 2021). Another interesting area of future research is the analysis of the joint implementation of various OC strategies, which complicates the implementation and the analysis (Gao et al., 2022). Additionally, the joint implementation of various OC strategies considering the forward and the reverse flows for a closed-loop approach would be an interesting area to explore in the future.

Future studies can also explore the last mile delivery decisions considering the delivery of products as well as return collections in an OCCLSC. Another interesting future research opportunity could be the network design for an OCSC from a closed-loop perspective. In order to incorporate channel integration while adopting a closed-loop perspective demands a revamped network design for an OC scenario. Analyzing this aspect could be an exciting subject for future investigation. An interesting area of future research is to explore the location inventory problem in an omnichannel closed-loop supply chain, thus combining the strategic and tactical decisions for an optimal OCSC. Order cancellations can have significant implications to inventory management and the literature attempts to capture this effect. However the studies are quite general and not specific to a particular type of retailer. E.g. an order cancellation for a grocery retailer has more severe consequences on the inventory management than an electronics retailers since the grocery products are perishable. Future studies can explore the effect of order cancellations on the inventory, logistics and warehousing strategies of various types of retailers. The assortment and bundling strategies in an OCSC remain largely unexplored, which can be analysed in future studies (Gao et al., 2022). Investigating the transshipment decisions for inventory considering returns is another area of future research.

5.2.2 Return Prediction, Return Avoidance, Incentive Schemes, and Reverse Flow Customer Steering

Given that accurate return predictions form vital inputs for strategic and tactical decisions of the managers, predicting returns and improving the accuracy results in substantial cost savings from returns management processes (Shang et al., 2020). Also, returns prediction is an essential component of product recovery management since it ensures the stability of the inventory control, remanufacturing processes, and reduction of uncertainty (Duong et al., 2022). Despite the vital role of returns prediction in returns management process, the literature lacks studies on this aspect. Future studies can explore the prediction of returns, prediction of the channel of return, quality of return, category of product return (e.g. return due to misfit, return due to dissatisfaction, return due to damage, etc.) and the probability of a fraudulent return. Consumer channel choice behavior can be predicted to a great degree based on the segment group allocated, and the segment groups are validated by demographic and endogenous factors (Gawor & Hoberg, 2019). Future studies should explore the role of advanced data analytics, Machine Learning (ML) and Artificial Intelligence (AI) that enable return prediction, inventory optimization and personalization of return policies (Stevenson & Rieck, 2024).

Future studies can also explore various returns avoidance measures, such as incentive schemes offered by the retailer to avoid/reduce returns. Future research can further advance our understanding of avoidance and reduction of fraudulent returns through incentive schemes. E.g. customers may be rewarded by the retailer through their rewards program for not returning a product (Wollenburg, Holzapfel et al., 2018). Retailers expect customers to return products to a low-cost-to-serve channel such as in-store return (Wollenburg, Holzapfel et al., 2018). In order to facilitate the return through the lowest cost-to-serve, the customer needs to be steered for the return. (Wollenburg, Holzapfel et al., 2018) and (Wollenburg et al., 2019) explore various reverse flow customer steering opportunities. However, more studies are required to explore the reverse flow customer steering opportunities that further deepen our understanding of the subject area. Future studies could explore the back-end requirements of reverse flow customer steering, such as software integration, ERP, or warehouse management (Wollenburg, Holzapfel et al., 2018). Another interesting area of future research is to quantify the implementation of various steering methods, the associated costs, and the actual benefit of implementing the strategies (Wollenburg, Holzapfel et al., 2018).

5.3 OC Returns Management Phase

5.3.1 Return Collection and Gatekeeping Decisions

In OC retailing, the customer has the flexibility to return the product through any channel. Since an online purchased product can be returned at any physical store, the retailers must decide on the returns gatekeeping decisions of whether to adopt centralized gatekeeping or decentralized gatekeeping. How retailer addresses this problem is a worthy topic for future research. Also, different types of retailers require different gatekeeping strategies. For e.g., grocery and fashion retailers might prefer decentralized gatekeeping due to the comparatively lower level of expertise needed for the evaluation of the returned product as compared to an electronics retailer who would prefer centralized gatekeeping due to the high level of expertise needed. This could be another avenue for future research. The retailer must also decide on returns collection decisions such as pickup of the return by the nearest physical store or the inhouse transportation or CEP carrier or outsourcing to a third party logistics service provider considering the forward flow in an OCCLSC.

5.3.2 Returns Inventory Management Decisions

It is crucial to manage the inventory of returned products optimally to ensure that the returns reach the desired destination instead of landfill. The systematic review highlights the scarcity of such studies. Future research can explore inventory rebalancing decisions such as partial/full retention of non-defective returned inventory at the physical store. Future research can also explore optimal store-to-store and DC-to-DC transshipment decisions of returned products in an OCCLSC. Future studies can explore innovative strategies to integrate the returned products back into the saleable inventory at a faster rate.

5.3.3 Recovery Process and Secondary Market Selling Decisions

As the non-defective returns are integrated into the saleable inventory, they need to undergo recovery processes. Optimal decisions on repairing, refurbishing, recycling, and selling in secondary markets still remain relatively underexplored in the literature. Future studies can explore the optimal decisions on selling the recovered returned products in the secondary markets. Another avenue for future research is to explore innovative and optimal strategies to recover defective returned products to ensure a Sustainable OCCLSC (SOCCLSC) and integrate the recovered returns back into the saleable inventory at a faster rate. Furthermore, the environmental and sustainability aspects of returns management in OCSC have been underexplored in the academic literature. The complexities of managing returns across the integrated channels pose substantial environmental challenges, such as increased carbon emissions, excessive consumption of resources, and lack of efficient strategies which can lead to increased waste and environmental degradation, underscoring the need for sustainable practices (Patra et al., 2024). Furthermore, the adoption of sustainable practices for returns management and reverse logistics in omnichannel scenarios is often hindered by barriers such as high cost, operational complexities, inadequate infrastructure, and lack of standardized processes (De Borba et al., 2020). Addressing these gaps through future research is crucial for developing sustainable, practical, and efficient strategies considering the unique complexities of omnichannel returns management.

6 Conclusion

Despite the complexities in returns management and reverse logistics and the additional intricacies involved in the case of OC, the literature on returns management and reverse logistics in OC is quite sparse. Considering the need to focus more on this aspect, this review aims to summarize the relevant literature. We provide a framework for the returns management and reverse logistics in OC and build upon the framework using the existing literature to provide a comprehensive overview of the current state of the literature. This research further categorized the literature into numerous categories and expanded on each theme to provide an in-depth outlook on each of the themes. It summarises the existing literature and proposes opportunities for future research in this field from a theoretical standpoint. It provides insight into critical returns management and reverse logistics decisions in OCSC for practitioners. In the domain of OCSC, this systematic review contributes to the literature by addressing the challenge of returns management and reverse logistics.

As OC gains more popularity among retailers and is widely adopted, the rate of returns is also increasing, thus posing severe challenges to the supply chains of the retailers. This study provides a comprehensive overview of the various challenges faced by retailers in returns management in OC and the potential improvements needed to overcome the challenges. It assists the managers in decision-making in various aspects of returns management and reverses logistics in OCSC.

A limitation of this study is the focus of the review on those articles that address the operational aspect of OC. Although the literature on returns in OC might be available in several other domains of research, such as marketing and human resources, that were not included in the sample and can be considered in future studies. The other limitation of this study is the inclusion of journal articles from Web of Science and Scopus databases in English language indexed in the ABDC journal rankings. While the authors of this paper prefer these databases for their pool of high-value journals, future studies can explore other databases such as EBSCO and Google Scholar. Also, we limit the search to articles in English and restricted only to journal articles. Future studies can also include conference proceedings and grey literature such as reports and book chapters. The findings of this research can help managers to improve their firms' efficiency levels by resizing their operational scale, strengthening their technology infrastructure and applications, and improving their discretionary allocation of resources.

6.1 Theoretical Implications

This study makes significant contributions to the existing body of literature on returns management and reverse logistics in OCCLSC by offering a systematic and structured understanding of the unique challenges in this area. The key contribution lies in the identification and in-depth exploration of several themes to categorize the existing literature based on phases of the omnichannel journey. This thematic analysis is crucial as it provides a novel framework for a holistic approach to understanding the multifaceted challenges posed by returns management and reverse logistics in OCCLSC. Moreover, the review highlights that the challenges to returns management and reverse logistics in an OCCLSC differ significantly from those faced in traditional offline e-commerce, and dual/multi-channel retail settings. Furthermore, the review provides a structured framework for future research directions that serves as a foundation for advancing scholarly inquiry in this area. By delineating key research themes and highlighting critical gaps, this study lays the groundwork for future theoretical advancements to address the evolving challenges in OCCLSC.

6.2 Practical Implications

This study provides valuable insights for practitioners transitioning to omnichannel by systematically identifying and analyzing critical challenges to returns management and reverse logistics in OCCLSC. The themes identified in this review inform the practitioners with a structured understanding of the recent developments and challenges across each phase of the omnichannel journey. By categorizing these challenges, the review enables practitioners to identify the specific pain-points and align their strategies with the unique demands of each phase. The review not only categorizes and synthesizes existing literature but also highlights proven strategies and emerging trends, allowing practitioners to adopt best practices for integrating forward and reverse flows. Furthermore, the study highlights the importance of a proactive approach by considering the reverse flow during each phase of the omnichannel journey in order to balance customer-centric return policies with operational efficiency in an omnichannel context.

References

Abdulla, H., Ketzenberg, M., & Abbey, J. D. (2019). Taking stock of consumer returns: A review and classification of the literature. *Journal of Operations Management, 65*(6), 560–605. https://doi.org/10.1002/joom.1047

Agatz, N. A. H., Fleischmann, M., & Van Nunen, J. A. E. E. (2008). E-fulfillment and multi-channel distribution—A review. *European Journal of Operational Research, 187*(2), 339–356. https://doi.org/10.1016/j.ejor.2007.04.024

Altug, M. S., & Aydinliyim, T. (2016). Counteracting strategic purchase deferrals: The impact of online retailers' return policy decisions. *Manufacturing & Service Operations Management, 18*(3), 376–392. https://doi.org/10.1287/msom.2015.0570

Ambilkar, P., Dohale, V., Gunasekaran, A., & Bilolikar, V. (2022). Product returns management: A comprehensive review and future research agenda. *International Journal of Production Research, 60*(12), 3920–3944. https://doi.org/10.1080/00207543.2021.1933645

Amini, M. M., Retzlaff-Roberts, D., & Bienstock, C. C. (2005). Designing a reverse logistics operation for short cycle time repair services. *International Journal of Production Economics, 96*(3), 367–380. https://doi.org/10.1016/j.ijpe.2004.05.010

Arslan, A. N., Klibi, W., & Montreuil, B. (2021). Distribution network deployment for omnichannel retailing. *European Journal of Operational Research, 294*(3), 1042–1058. https://doi.org/10.1016/j.ejor.2020.04.016

Beck, N., & Rygl, D. (2015). Categorization of multiple channel retailing in Multi-, Cross-, and Omni-Channel Retailing for retailers and retailing. *Journal of Retailing and Consumer Services, 27*, 170–178. https://doi.org/10.1016/j.jretconser.2015.08.001

Bell, D. R., Gallino, S., & Moreno, A. (2018). Offline showrooms in omnichannel retail: Demand and operational benefits. *Management Science, 64*(4), 1629–1651. https://doi.org/10.1287/mnsc.2016.2684

Bernon, M., Cullen, J., & Gorst, J. (2016). Online retail returns management: Integration within an omni-channel distribution context. *International Journal of Physical Distribution & Logistics Management, 46*(6/7), 584–605. https://doi.org/10.1108/IJPDLM-01-2015-0010

Bijmolt, T. H. A., Broekhuis, M., De Leeuw, S., Hirche, C., Rooderkerk, R. P., Sousa, R., & Zhu, S. X. (2021). Challenges at the marketing–operations interface in omni-channel retail environments. *Journal of Business Research, 122*, 864–874. https://doi.org/10.1016/j.jbusres.2019.11.034

Brynjolfsson, E., Hu, Y. J., & Rahman, M. S. (2013). As technology blurs the distinctions between physical and online retailing, retailers and their supply-chain partners will need to rethink their competitive strategies. *DIGITAL TRANSFORMATION*.

Cai, Y.-J., & Lo, C. K. Y. (2020). Omni-channel management in the new retailing era: A systematic review and future research agenda. *International Journal of Production Economics, 229*, Article 107729. https://doi.org/10.1016/j.ijpe.2020.107729

De Borba, J. L. G., Magalhães, M. R. D., Filgueiras, R. S., & Bouzon, M. (2020). Barriers in omnichannel retailing returns: A conceptual framework. *International Journal of Retail & Distribution Management, 49*(1), 121–143. https://doi.org/10.1108/IJRDM-04-2020-0140

De Leeuw, S., Minguela-Rata, B., Sabet, E., Boter, J., & Sigurðardóttir, R. (2016). Trade-offs in managing commercial consumer returns for online apparel retail. *International Journal of Operations & Production Management, 36*(6), 710–731. https://doi.org/10.1108/IJOPM-01-2015-0010

Duong, Q. H., Zhou, L., Meng, M., Nguyen, T. V., Ieromonachou, P., & Nguyen, D. T. (2022). Understanding product returns: A systematic literature review using machine learning and bibliometric analysis. *International Journal of Production Economics, 243*, Article 108340. https://doi.org/10.1016/j.ijpe.2021.108340

Durach, C. F., Kembro, J., & Wieland, A. (2017). A New Paradigm for systematic literature reviews in supply chain management. *Journal of Supply Chain Management, 53*(4), 67–85. https://doi.org/10.1111/jscm.12145

Gao, F., Agrawal, V. V., & Cui, S. (2022). The effect of multichannel and omnichannel retailing on physical stores. *Management Science, 68*(2), 809–826. https://doi.org/10.1287/mnsc.2021.3968

Gao, F., & Su, X. (2017a). Omnichannel retail operations with buy-online-and-pick-up-in-store. *Management Science, 63*(8), 2478–2492. https://doi.org/10.1287/mnsc.2016.2473

Gao, F., & Su, X. (2017b). Online and offline information for omnichannel retailing. *Manufacturing & Service Operations Management, 19*(1), 84–98. https://doi.org/10.1287/msom.2016.0593

Gawor, T., & Hoberg, K. (2019). Customers' valuation of time and convenience in e-fulfillment. *International Journal of Physical Distribution & Logistics Management, 49*(1), 75–98. https://doi.org/10.1108/IJPDLM-09-2017-0275

Gerea, C., Gonzalez-Lopez, F., & Herskovic, V. (2021). Omnichannel customer experience and management: An integrative review and research agenda. *Sustainability, 13*(5), 2824. https://doi.org/10.3390/su13052824

Goedhart, J., Haijema, R., & Akkerman, R. (2023). Modelling the influence of returns for an omni-channel retailer. *European Journal of Operational Research, 306*(3), 1248–1263. https://doi.org/10.1016/j.ejor.2022.08.021

Govindan, K., & Bouzon, M. (2018). From a literature review to a multi-perspective framework for reverse logistics barriers and drivers. *Journal of Cleaner Production, 187*, 318–337. https://doi.org/10.1016/j.jclepro.2018.03.040

Grewal, D., Iyer, G. R., & Levy, M. (2004). Internet retailing: Enablers, limiters and market consequences. *Journal of Business Research, 57*(7), 703–713. https://doi.org/10.1016/S0148-2963(02)00348-X

Guerrero-Lorente, J., Gabor, A. F., & Ponce-Cueto, E. (2020). Omnichannel logistics network design with integrated customer preference for deliveries and returns. *Computers & Industrial Engineering, 144*, Article 106433. https://doi.org/10.1016/j.cie.2020.106433

Guerrero-Lorente, J., Ponce-Cueto, E., & Blanco, E. E. (2017). A model that integrates direct and reverse flows in omnichannel logistics networks. In M. Amorim, C. Ferreira, M. Vieira Junior, & C. Prado (Eds.), *Engineering systems and networks* (pp. 89–97). Springer International Publishing. https://doi.org/10.1007/978-3-319-45748-2_10

Hao, J., Jr, R. G. R., Morgan, T. R., & Slazinik, I. M. (2024). The BORIS experience: Evaluating omnichannel returns and repurchase intention. *International Journal of Physical Distribution & Logistics Management, 54*(11), 44–81. https://doi.org/10.1108/IJPDLM-09-2023-0357

He, Y., Xu, Q., & Wu, P. (2020). Omnichannel retail operations with refurbished consumer returns. *International Journal of Production Research, 58*(1), 271–290. https://doi.org/10.1080/00207543.2019.1629672

Hochrein, S., & Glock, C. H. (2012). Systematic literature reviews in purchasing and supply management research: A tertiary study. *International Journal of Integrated Supply Management, 7*(4), 215–245. https://doi.org/10.1504/IJISM.2012.052773

Hole, Y., Pawar, Ms. S., & Khedkar, E. B. (2019). Omni channel retailing: An opportunity and challenges in the Indian Market. *Journal of Physics: Conference Series, 1362*(1), 012121. https://doi.org/10.1088/1742-6596/1362/1/012121

Hübner, A., Hense, J., & Dethlefs, C. (2022). The revival of retail stores via omnichannel operations: A literature review and research framework. *European Journal of Operational Research, 302*(3), 799–818. https://doi.org/10.1016/j.ejor.2021.12.021

Hübner, A., Holzapfel, A., & Kuhn, H. (2015). Operations management in multi-channel retailing: An exploratory study. *Operations Management Research, 8*(3–4), 84–100. https://doi.org/10.1007/s12063-015-0101-9

Hübner, A., Wollenburg, J., & Holzapfel, A. (2016). Retail logistics in the transition from multi-channel to omni-channel. *International Journal of Physical Distribution & Logistics Management, 46*(6/7), 562–583. https://doi.org/10.1108/IJPDLM-08-2015-0179

Hwang, E. H., Nageswaran, L., & Cho, S.-H. (2022). Value of online–off-line return partnership to off-line retailers. *Manufacturing & Service Operations Management, 24*(3), 1630–1649. https://doi.org/10.1287/msom.2021.1026

Jena, S. K., & Meena, P. (2022). Shopping in the omnichannel supply chain under price competition and product return. *Journal of Retailing and Consumer Services, 65*, Article 102848. https://doi.org/10.1016/j.jretconser.2021.102848

Jin, D., Caliskan-Demirag, O., Chen, F. (Youhua), & Huang, M. (2020). Omnichannel retailers' return policy strategies in the presence of competition. *International Journal of Production Economics, 225*, 107595. https://doi.org/10.1016/j.ijpe.2019.107595

Jin, D., & Huang, M. (2021). Competing e-tailers' adoption strategies of buy-online-and-return-in-store service. *Electronic Commerce Research and Applications, 47*, Article 101047. https://doi.org/10.1016/j.elerap.2021.101047

Jin, M., Li, G., & Cheng, T. C. E. (2018). Buy online and pick up in-store: Design of the service area. *European Journal of Operational Research, 268*(2), 613–623. https://doi.org/10.1016/j.ejor.2018.02.002

Jing, B. (2018). Showrooming and Webrooming: information externalities between online and offline sellers. *Marketing Science, 37*(3), 469–483. https://doi.org/10.1287/mksc.2018.1084

Kembro, J. H., Norrman, A., & Eriksson, E. (2018). Adapting warehouse operations and design to omni-channel logistics: A literature review and research agenda. *International Journal of Physical Distribution & Logistics Management, 48*(9), 890–912. https://doi.org/10.1108/IJPDLM-01-2017-0052

Li, R. (2020). Reinvent retail supply chain: Ship-from-store-to-store. *Production and Operations Management, 29*(8), 1825–1836. https://doi.org/10.1111/poms.13195

Li, Y. (2021). Multiperiod dynamic pricing and inventory control decisions for an omnichannel BOPS retailer with reference price effects. *Mathematical Problems in Engineering, 2021*(1), 9650805. https://doi.org/10.1155/2021/9650805

Li, Z., Yang, W., & Chen, X. (2021). Omnichannel inventory models accounting for buy-online–return-to-store service and random demand. *Soft Computing, 25*(17), 11691–11710. https://doi.org/10.1007/s00500-021-06045-0

Liberati, A., Altman, D. G., Tetzlaff, J., Mulrow, C., Gøtzsche, P. C., Ioannidis, J. P. A., Clarke, M., Devereaux, P. J., Kleijnen, J., & Moher, D. (2009). The PRISMA statement for reporting systematic reviews and meta-analyses of studies that evaluate health care interventions: Explanation and elaboration. *PLoS Medicine, 6*(7), Article e1000100. https://doi.org/10.1371/journal.pmed.1000100

Liu, J., & Xu, Q. (2020). Joint decision on pricing and ordering for omnichannel BOPS retailers: Considering online returns. *Sustainability, 12*(4), 1539. https://doi.org/10.3390/su12041539

Liu, L., Feng, L., Xu, B., & Deng, W. (2020). Operation strategies for an omni-channel supply chain: Who is better off taking on the online channel and offline service? *Electronic Commerce Research and Applications, 39*, Article 100918. https://doi.org/10.1016/j.elerap.2019.100918

Mandal, P., Basu, P., & Saha, K. (2021). Forays into omnichannel: An online retailer's strategies for managing product returns. *European Journal of Operational Research, 292*(2), 633–651. https://doi.org/10.1016/j.ejor.2020.10.042

Melacini, M., Perotti, S., Rasini, M., & Tappia, E. (2018). E-fulfilment and distribution in omni-channel retailing: A systematic literature review. *International Journal of Physical Distribution & Logistics Management, 48*(4), 391–414. https://doi.org/10.1108/IJPDLM-02-2017-0101

Muir, W. A., Griffis, S. E., & Whipple, J. M. (2019). A simulation model of multi-echelon retail inventory with cross-channel product returns. *Journal of Business Logistics, 40*(4), 322–338. https://doi.org/10.1111/jbl.12209

Nageswaran, L., Cho, S.-H., & Scheller-Wolf, A. (2020). Consumer return policies in omnichannel operations. *Management Science, 66*(12), 5558–5575. https://doi.org/10.1287/mnsc.2019.3492

Ofek, E., Katona, Z., & Sarvary, M. (2011). "Bricks and clicks": The impact of product returns on the strategies of multichannel retailers. *Marketing Science, 30*(1), 42–60. https://doi.org/10.1287/mksc.1100.0588

Orús, C., Gurrea, R., & Ibáñez-Sánchez, S. (2019). The impact of consumers' positive online recommendations on the omnichannel webrooming experience. *Spanish Journal of Marketing - ESIC, 23*(3), 397–414. https://doi.org/10.1108/SJME-08-2019-0067

Patra, P., Panja, S., & Mondal, S. K. (2024). Mitigating consumer returns through omnichannel retail operations: An EOQ model under stochastic demand. *Journal of Systems Science and Systems Engineering*. https://doi.org/10.1007/s11518-024-5617-9

Raza, S. A., & Govindaluri, S. M. (2021). Omni-channel retailing in supply chains: A systematic literature review. *Benchmarking: An International Journal, 28*(9), 2605–2635. https://doi.org/10.1108/BIJ-10-2020-0547

Rigby, D. (2011, December 1). The future of shopping. *Harvard Business Review*. https://hbr.org/2011/12/the-future-of-shopping

Rooderkerk, R. P., & Kök, A. G. (2019). Omnichannel assortment planning. In S. Gallino & A. Moreno (Eds.), *Operations in an omnichannel world* (pp. 51–86). Springer International Publishing. https://doi.org/10.1007/978-3-030-20119-7_4

Saghiri, S., Wilding, R., Mena, C., & Bourlakis, M. (2017). Toward a three-dimensional framework for omni-channel. *Journal of Business Research, 77*, 53–67. https://doi.org/10.1016/j.jbusres.2017.03.025

Serkan Akturk, M., Ketzenberg, M., & Heim, G. R. (2018). Assessing impacts of introducing ship-to-store service on sales and returns in omnichannel retailing: A data analytics study. *Journal of Operations Management, 61*(1), 15–45. https://doi.org/10.1016/j.jom.2018.06.004

Shang, G., McKie, E. C., Ferguson, M. E., & Galbreth, M. R. (2020). Using transactions data to improve consumer returns forecasting. *Journal of Operations Management, 66*(3), 326–348. https://doi.org/10.1002/joom.1071

Shi, X., Dong, C., & Cheng, T. C. E. (2018). Does the buy-online-and-pick-up-in-store strategy with pre-orders benefit a retailer with the consideration of returns? *International Journal of Production Economics, 206*, 134–145. https://doi.org/10.1016/j.ijpe.2018.09.030

Snyder, H. (2019). Literature review as a research methodology: An overview and guidelines. *Journal of Business Research, 104*, 333–339. https://doi.org/10.1016/j.jbusres.2019.07.039

Stevenson, A. B., & Rieck, J. (2024). Investigating returns management across e-commerce sectors and countries: Trends, perspectives, and future research. *Logistics, 8*(3), Article 3. https://doi.org/10.3390/logistics8030082

Sun, H., & Tian, Y. (2018). Using improved genetic algorithm under uncertain circumstance of site selection of O2O customer returns. *International Journal of Data Analysis Techniques and Strategies, 10*(3), 241–256. https://doi.org/10.1504/IJDATS.2018.094129

Taylor, D., Brockhaus, S., Knemeyer, A. M., & Murphy, P. (2019). Omnichannel fulfillment strategies: Defining the concept and building an agenda for future inquiry. *The International Journal of Logistics Management, ahead-of-print* (ahead-of-print). https://doi.org/10.1108/IJLM-09-2018-0223

Thaichon, P., Quach, S., Barari, M., & Nguyen, M. (2024). Exploring the role of omnichannel retailing technologies: Future research directions. *Australasian Marketing Journal, 32*(2), 162–177. https://doi.org/10.1177/14413582231167664

Tranfield, D., Denyer, D., & Smart, P. (2003). Towards a methodology for developing evidence-informed management knowledge by means of systematic review. *British Journal of Management, 14*(3), 207–222. https://doi.org/10.1111/1467-8551.00375

Verhoef, P. C., Kannan, P. K., & Inman, J. J. (2015). From multi-channel retailing to omni-channel retailing. *Journal of Retailing, 91*(2), 174–181. https://doi.org/10.1016/j.jretai.2015.02.005

Wollenburg, J., Holzapfel, A., & Hübner, A. (2019). *Omni-channel customer management processes in retail: An exploratory study on fulfillment-related options* (7th ed.). Bundesvereinigung Logistik (BVL) e.V. https://doi.org/10.23773/2019_7

Wollenburg, J., Holzapfel, A., Hübner, A., & Kuhn, H. (2018). Configuring retail fulfillment processes for omni-channel customer steering. *International Journal of Electronic Commerce, 22*(4), 540–575. https://doi.org/10.1080/10864415.2018.1485085

Wollenburg, J., Hübner, A., Kuhn, H., & Trautrims, A. (2018b). From bricks-and-mortar to bricks-and-clicks: Logistics networks in omni-channel grocery retailing. *International Journal of Physical Distribution & Logistics Management, 48*(4), 415–438. https://doi.org/10.1108/IJPDLM-10-2016-0290

Wu, D., & Lu, X. (2023). Optimal return policy of competitive retailers' pre-sale products based on strategic consumer behavior. *Sustainability, 15*(2), Article 2. https://doi.org/10.3390/su15021341

Xu, X., & Jackson, J. E. (2019). Investigating the influential factors of return channel loyalty in omni-channel retailing. *International Journal of Production Economics, 216*, 118–132. https://doi.org/10.1016/j.ijpe.2019.03.011

Yang, G., & Ji, G. (2022). The impact of cross-selling on managing consumer returns in omnichannel operations. *Omega, 111*, Article 102665. https://doi.org/10.1016/j.omega.2022.102665

Yang, L., Li, X., Xia, Y., & Aneja, Y. P. (2023). Returns operations in omnichannel retailing with buy-online-and-return-to-store. *Omega, 119*, Article 102874. https://doi.org/10.1016/j.omega.2023.102874

Zhang, J., Farris, P. W., Irvin, J. W., Kushwaha, T., Steenburgh, T. J., & Weitz, B. A. (2010). Crafting integrated multichannel retailing strategies. *Journal of Interactive Marketing, 24*(2), 168–180. https://doi.org/10.1016/j.intmar.2010.02.002

Zhang, J., Li, H., Yan, R., & Johnston, C. (2017). Examining the signaling effect of e-tailers' return policies. *Journal of Computer Information Systems, 57*(3), 191–200. https://doi.org/10.1080/08874417.2016.1183989

Zhang, J., Xu, Q., & He, Y. (2018). Omnichannel retail operations with consumer returns and order cancellation. *Transportation Research Part E: Logistics and Transportation Review, 118*, 308–324. https://doi.org/10.1016/j.tre.2018.08.006

Optimal Consumer Targeting Utilizing Consumer Returns and Bad Debt: A Discrete Choice Application

Nian Wang, Joseph Pancras, Hongju Liu, and Malcolm Houtz

Abstract Dealing with 'demon customers' (Selden, L., & Colvin, G. (2003). Angel customers & demon customers: Discover which is which and turbo-charge your stock. Penguin.) consumes substantial resources, and leads to continual inefficiency and sub-optimal profits for firms. Undesirable behavior exhibited by such customers can manifest in two major types of behavior—unwarranted returns and bad debt. Academic research has largely focused on the effect of returns, detailing conditions under which return behavior can be profitable (Petersen and Kumar, Journal of Marketing 73:35–51, 2009). Less is known about the impact of bad debt, even though predicting bad debt is an important marketing activity for firms in the direct response industry. Firms in this industry market products to a substantial segment of customers with limited access to credit card debt. For such customers transaction specific 'credit rating' is not predetermined (as in credit cards), but is rather an integral part of the targeting process through the 'bill me later' or 'buy now pay later' (BNPL) payment mechanism. Thus, little is also known about the following research question: what is the combined impact of both return and bad debt behaviors on firm profitability? We study this important question with the help of a unique dataset from a co-operative database, and find that direct response firms may benefit from reconsidering their return policies by factoring in customer bad debt propensities and costs. Our model and results continue to be relevant in the post-pandemic era of the 2020s with the rise of BNPL platforms such as Klarna, Affirm.

N. Wang
University of Connecticut School of Business (PhD 2018), Storrs, CT, USA

J. Pancras (✉)
University of Connecticut School of Business, Storrs, CT, USA
e-mail: Joseph.Pancras@uconn.edu

H. Liu
Guanghua School of Management, Peking University, Beijing, China
e-mail: hliu@gsm.pku.edu.cn

M. Houtz
Alliant Inc, Brewster, NY, USA
e-mail: mhoutz@alliantdata.com

B. Basu et al. (eds.), *Data Analytics for Decision Making towards Business Excellence*, Uncertainty and Operations Research, https://doi.org/10.1007/978-981-95-2429-7_5

Keywords Product return · Bad debt · Direct mail · Targeted Marketing

1 Introduction

Firms targeting customers trade off maximizing response with minimizing costs due to undesirable customers. Increasing credit card usage by US consumers (Durkin, 2000) implies that for most commonly studied firms the costs of targeting undesirables are restricted to the costs of returns, since the risk of non-payment is assumed by credit card companies. However, there is a substantial and largely understudied industry called the 'direct response' industry (also called 'direct marketing' or 'direct mail') that largely serves the needs of customers who may have limited or no access to credit card debt. Many firms in this industry use a 'bill me later' payment mechanism that does not require a credit card. The rise of "buy now pay later" payment platforms such as Klarna, Affirm and Afterpay during and after the 2020 pandemic points to the continued relevance of this sector (Kumar et al., 2024; Stavins, 2024). Since customers most often pay by check after receiving the product, this introduces an additional risk in the targeting decision of costs that could be incurred due to non-payment by customers, termed risk of 'bad debt' or 'default'. Thus, firms in the direct response industry grapple with dual issues of minimizing returns and bad debt in crafting their targeting algorithms.

Direct marketing continues to be important to marketers in the 2020s, as shown by this description from the industry publication in 2021 (Vatne, 2021):

> Amid the entire digital revolution in advertising, direct mail has remained a reliable constant for performance-minded marketers. Direct mail has been around for generations, and as a result has garnered much less attention than other channels. But look under the hood at many fast-growing DTC brands, and you'll find a steady commitment to growing through catalogs and mailers, all of which are, with the help of data, still very much consistent with an outcomes-based approach. As many DTCs hit saturation points in social and search, they have turned to direct mail as a proven way to find, engage, and convert new customers. DTCs are just the most notable recent example; direct mail remains a lifeline for many nonprofits and mission-driven organizations in the hunt for donors, subscribers, and members.

Returns have been allowed by direct marketers because customers are more likely to order the products when they have a return option (e.g. Anderson et al., 2009). Firms may penalize excessive return behavior through various disincentives such as a penalty for returned products. However, managing product returns is complex since there is a non-linear (inverted U) relationship between the number of product returns and a customer's profitability to a firm. There is a robust understanding of the effect of consumer returns on retailer profitability in the marketing academic literature (Petersen & Kumar, 2009).

By contrast bad debt has been relatively less studied in the marketing literature. Liu et al. (2015) show that firms can substantially improve profits by incorporating information on customer bad debt behaviors in their targeting policies. When examining the undesirable behaviors of returns and bad debt in tandem, the question that is of paramount importance from the direct response firm's point of view is the relative magnitude of the costs of return and bad debt. In many cases the answer may not be as obvious as it may seem at first glance. Here bad debt refers to the scenario where the consumer defaults on the credit that they have been extended and refuse to pay the amount owed for the product that has already been delivered to them. In some situations the bad debt option can cost less than the return option for the firm! This may be due to high labor costs related to operations such as product handling or restocking costs, which may end up exceeding the cost of the product itself. This raises the counter-intuitive research question that is addressed in this paper: what would be the impact of a firm's policy to not allow product returns[1]?

In this paper we measure the differential impact of bad debt information on the profitability of a firm's targeted marketing policies. We first calibrate the impact of return information, largely replicating the relationships reported in earlier research (Petersen & Kumar, 2009). Then we measure the impact of adding bad debt information incrementally demonstrating its value in predicting returns, then providing evidence for the need to explicitly incorporate bad debt prediction into the targeting algorithm.

The major contributions of this paper to the literature on returns and customer targeting are as follows. To our knowledge our study is the first to find a positive correlation between customer preferences for the return option and those for the bad debt option, which implies that the bad debt option cannot be ignored in the study of product returns. We also find that removing the return option may actually increase the firm's profit: our framework suggests that the firm should target customers with more payments or more returns in the past when the product return is not allowed. Substantively this study is also, to our knowledge, one of the first to study unique marketing issues related to the substantial segment of customers (of up to 28% of the US population) who have limited or no access to credit card debt.

In Sect. 2, we describe the related literature on product returns and targeting in direct marketing. In Sect. 3, we construct the discrete choice modeling framework and compare the possible models. In Sect. 4, we estimate the models using data obtained from a Spanish book company. In Sect. 5, we use the estimates in Sect. 4 to perform a simulation study of removing the return option, and discuss the optimal targeting scheme when the return option is removed. We conclude in Sect. 7.

[1] Note that this would allow customers to retain products without payment (become 'bad debts') if they choose to do so. While somewhat counterintuitive, this may be a plausible alternative for firms for whom the cost of collecting bad debt is greater than the cost of the product.

2 Industry Background, Theory and Literature Review

2.1 The Response/Screening Tradeoff in the Direct Response Industry

Firms employing targeted marketing face a key tradeoff in terms of credit screening and customer response. With the widespread use of credit cards in both online and offline commerce, credit screening has largely been outsourced to the banks and companies that issue credit cards. However, the proportion of customers who do not own a credit card due to low income or past credit issues is still about 28% of US consumers (Holmes, 2014),[2] which amounts to 65.8 million consumers. Companies that serve this sizeable consumer segment have to find ways to screen out consumers who are bad debts among these from those who will pay.

In addition to this structural targeting issue, there is likely an under-researched propensity of customers to respond when they are invited to 'buy now and pay later', which could be perceived as being given a 'pass' on screening. This is likely to exist not only among the 65.8 million customers who do not own a credit card, but also probably a sizeable proportion among the 33% who own 1–2 credit cards and who may have limited credit. The 'bill me later' payment mechanism is likely to cue impulse purchase (Gilbride et al., 2015; Rook, 1987) though we are not aware of research that explicitly studies the mechanism underlying this tradeoff. However, the tradeoff exists, as is evidenced by the existence of the sizeable 'direct response industry' that is described in the following section. While tracing this mechanism is, we believe, a fruitful area for future research, in this paper we focus more on the managerial implications of this tradeoff in a real field study using appropriate econometric methods to accurately document the response of customers to targeted marketing in a direct response context. We also identify bad debt prediction and management through refining targeting algorithms as one unique feature of this segment.

While the terms 'catalog marketing' and 'direct response' are used in similar contexts, catalog marketing often requires payment with a credit card or a check upfront. This is to be distinguished from the direct response industry, where the payment terms are often on a 'bill me later' basis. Essentially here the credit rating component of the commercial transaction is not outsourced, but instead is made an integral part of the targeting process. In other words, targeting has to be optimized to not just maximize response, but also to minimize the possibility of bad debt.

This paper examines bad-debt and return policies in the context of the direct response industry, which approached $2 billion in revenues in 2014 (Statista, 2013). We focus on a subset of this industry known as the 'bill-me-later' market.

[2] This proportion has come down to about 15.7% in 2023, according to the 2023 FDIC Survey (https://www.fdic.gov/household-survey/2023-fdic-national-survey-unbanked-and-underbanked-households-report). The same survey reports that almost 4% of households used Buy Now Pay Later (BNPL) in 2023.

These are direct response product or service offers which do not require payment prior to order shipment. Bill-me-later offers expose the marketer to the risk of consumer bad-debt, but simultaneously increases response rates. Many well-known companies promote bill-me-later offers. MBI (collectibles, jewelry, gifts, Bradford Exchange (collectibles, jewelry, gifts), Boardroom (newsletters) and Publishers Clearing House (magazines, merchandise) have combined revenue approaching $1.5 billion.[3] Furthermore, many of the 7,000 magazines published in the U.S. also offer bill-me-later options. The bill-me-later business model has also morphed into major applications outside of the direct response industry as well, probably to take advantage of its ability to increase response rates. A striking example of this was the 2008 purchase by the leading online payments system company Paypal of 'Bill Me Later', a company that extends credit on a transaction by transaction basis for almost $1 billion (Bruene, 2008).

2.2 *Product Returns and Bad Debts*

Two questions have dominated the research on product returns: (1) Should the firm offer a return option? (2) Should the firm charge for the product return? Firms generally offer a return option because customers are less likely to order products if the return option is not allowed (Anderson et al., 2009). Firms also allow product returns from retailers if there is a viable used market for the product, where customers have a high willingness to pay for used products (Gümüs et al., 2013). Furthermore, product return behavior can be managed to increase lifetime value of customers; Petersen and Kumar (2009) find that customer value is maximized at an intermediate level of returns, rather than either a complete absence or an extremely high level of returns.

As product returns can be costly, many retailers charge a restocking fee for the product return. The return penalty may be more severe when product returns are salvaged by a channel member because the retailer misses out on the generous refund from the manufacturer when the retailer salvages returned units without the help of the manufacturer (Shulman et al., 2010). However, things are different for online retailers. They should either institute a policy of free product returns or at least examine their customer data to determine their customers' responses to return fees (Bower and Maxham III, 2012). This is because customers who paid for their product returns decreased their post-return spending at that retailer; and those who had free returns increased their post-return spending.

The extant literature on returns does not consider the bad debt option when studying returns. Liu et al. (2015) have shown that the firm can significantly increase its profit after incorporating bad debt behaviors. However, they do not address the

[3] Industry sources and online links were accessed to obtain this estimate. See the following links: https://www.statista.com/topics/1265/magazines/, http://www.mbi-inc.com/, https://www.google.com/?gws_rd=ssl#q=publishers+clearing+house+revenue, all accessed November 4, 2016.

issue whether bad debt and return behavior are related, or trace out consequent implications for firm targeting. An intriguing possibility arising from this perspective of considering bad debt and returns as possibly related behaviors is whether the firm may actually prefer bad debt to product return when the cost of return is greater. This could occur if the firm were to find it optimal to not allow product returns under certain conditions, which would provide a counterpoint to the recommendations of the previous literature on product returns.

2.3 *Non-Random Targeting in Catalog Mailing*

Past responses of customers to firm targeting reveal their preferences, and it is common practice in industry for firms to target customers based on their responsiveness to past targeting by firms. While this is intuitive, the practice can cause bias in customer response models that do not account for this non-random firm targeting behavior. Firms employing customer response models to make targeting decisions thus have to correct for their own targeting behavior in the past. In other words, before the firm decides to send a mailing to a certain customer, it should consider how many mailings have been sent to that customer. Ignoring the targeted histories in the modeling process for individual level targeting biases the parameter estimates (Dong et al., 2009). Several papers provide possible reasons for these biases. For example, customers have an adverse reaction to direct mail activity because of heavy direct mail activity by the firm in the past (Hartmann et al., 2011). But the effect can be temporary. The negative influence of extra mailings on future responding decisions dies out after one year (van Diepen, Donkers, & Franses, 2009a). When the firm sends mail intermittently and achieves its full impact instead of continually mailing and diminishing their effectiveness, the total response is higher (Ansari et al., 2008). Firms can increase the profits when targeting over multiple periods as opposed to a single period (Gonul, Frenkel Ter Hofstede, 2006). Zhang et al. (2022) show that returns of "deep" products (products that require physical inspection for choice evaluation) predict whether customers transition to a higher value state. In contrast to the earlier literature, which corrects for bias in purchase decisions only, our study shows the effect of such bias in two additional response options, return and bad debt customer response.

Most of the previous literature suggests that the firm should target customers with higher probability of response. For example Bodapati (2008) suggests that the firm should target customers whose purchase probabilities will increase if they receive the targeted mailings. We use sensitivity of purchase probabilities and measure the customers' response probabilities under the scenario where product returns are not allowed. A summary of major contributions of this study with respect to the extant literature discussed above is given in Table 1.

Table 1 Summary of literature

Paper	Main question	Main findings	Type	Choice alternatives
Anderson, Hansen, Simester (2009)	Return option value	The customers are less likely to order the products if the return option is removed	Empirical	No response, buy, return
Gümüs et al. (2013)	Return policy and customer valuation of used products	The higher the customer valuation of used products is, the more likely the company would offer a return option to the retailer	Analytical	No response, buy, return
Petersen and Kumar (2009)	Return behavior and CLV	As returns increase, the lifetime value of a customer increases first and then starts decreasing	Empirical	No response, buy, return
Shulman, Coughlan, and Savaskan (2010)	Return penalty	The return penalty may be more severe when returns are salvaged by a channel member who derives greater value from a returned unit	Analytical	No response, buy, return
Bower and Maxham III (2012)	Free returns vs. fee returns	Customers who paid for their own return decreased their postreturn spending at that retailer, and vice versa	Empirical	No response, buy, return
Hartmann et al. (2011)	Causal marketing mix effects	Customers have adverse reactions to direct mail activity because of heavy direct mail activity by the firm in the past	Empirical	No response, response
van Diepen, Donkers, and Franses (2009a)	Dynamic effects and competitive effects	Each extra mailing the same charity sends negatively affects the future responding decision; but the effects die out in one year	Empirical	No response, response
Gonul, Frenkel Ter Hofstede (2006)	Mailing strategies	The expected profits are higher when a longer time horizon is employed than a single-period horizon	Empirical	No response, response
Ansari et al. (2008)	Customer channel migration	The total response can be higher by sending e-mails intermittently than by continually e-mailing	Empirical	No response, response

(continued)

Table 1 (continued)

Paper	Main question	Main findings	Type	Choice alternatives
Bodapati (2008)	Targeting with purchase data	The targeting should be based on the sensitivity of purchase probabilities instead of the purchase probabilities	Empirical	No response, response
Dong et al. (2009)	Targeting in the presence of firm strategic behavior	Ignoring firm strategic behavior in the modeling process biases the parameter estimates	Empirical	No response, response
Zhang et al. (2022)	Role of physical store purchase and returns on customer value	Returns of "deep" products (that require physical inspection for evaluation) transitions customers to a higher value state	Empirical	Physical/online channel choice, purchase amount, and interpurchase time
This Study	Targeting customers with return-bad-debt tradeoff	Historical bad debts predict return probabilities; firms can increase profit by dropping the return option	Empirical	No response, paid, return, bad debt

3 Data and Models

The data is obtained from a co-operative database contributor which sells Spanish books by direct mail,[4] and contains customers' historical transaction information such as the number of historical cancels/returns, the number of historical payments, the total return count and so on. The company selected certain customers to send direct mail promotions on six dates.[5] There are 140,208 customers and 200,779 observations in total.

We adopt a two-pronged modeling strategy to study the research question about the interplay between return and bad debt modeling in the targeting decision of firms. In the first we model as *continuous* dependent variables the responses of customers to firm targeting in terms of the number of orders placed, the number of returns and the number of bad debts. In the second approach we model the probabilities of incidence of each of these *discrete* responses. This approach has the advantage of tractability as well as flexibility to account for both the continuous and discrete responses of the consumer to direct marketing efforts of the firm. We report major findings from each approach below.[6] Below we develop a conceptual framework and utilize the frame work to select predictors for the continuous and discrete response models which are developed in subsequent sections.

---

[4] The book is a single book called a 'hints and tips type book', written in Spanish. The type of promotion is often called a 'solo offer', where just one item promoted in the mail piece.

[5] These are termed archive dates by the company.

[6] More details of the analyses are available on request.

3.1 Continuous Response Model

We model customer responses to firm targeting using a seemingly unrelated regression (SUR) approach. The simultaneous equations comprising this method are:

$$\begin{aligned} Order_{it} &= \alpha_0 + \alpha_1 TON_{it} + \alpha_2 \left(TON_{it}^2\right)^{1/3} + \alpha_3 HON_{it} + \alpha_4 \left(HON_{it}^2\right)^{1/3} \\ &\quad + \alpha_5 ORN_{it} + \alpha_6 \left(ORN_{it}^2\right)^{1/3} + \alpha_7 TBN_{it} + \alpha_8 \left(TBN_{it}^2\right)^{1/3} + \beta' Z_{OR} + \varepsilon_{OR} \end{aligned} \tag{1}$$

$$\begin{aligned} Return_{it} &= \delta_0 + \delta_1 ORN_{it} + \delta_2 \left(ORN_{it}^2\right)^{1/3} + \delta_3 TON_{it} + \delta_4 \left(TON_{it}^2\right)^{1/3} + \delta_5 HON_{it} + \\ &\quad \delta_6 \left(HON_{it}^2\right)^{1/3} + \delta_7 TBN_{it} + \delta_8 \left(TBN_{it}^2\right)^{1/3} + \lambda' Z_{RE} + \varepsilon_{RE} \end{aligned} \tag{2}$$

$$\begin{aligned} Baddebt_{it} &= \omega_0 + \omega_1 TBN_{it} + \omega_2 \left(TBN_{it}^2\right)^{1/3} + \omega_3 ORN_{it} + \omega_4 \left(ORN_{it}^2\right)^{1/3} + \omega_5 TON_{it} + \\ &\quad \omega_6 \left(TON_{it}^2\right)^{1/3} + \omega_7 HON_{it} + \omega_8 \left(HON_{it}^2\right)^{1/3} + \phi' Z_{BD} + \varepsilon_{BD} \end{aligned} \tag{3}$$

Where $(\varepsilon_{OR}, \varepsilon_{RE}, \varepsilon_{BD})' \sim \text{iidN}(0, \Omega'')$.

Descriptions and descriptive statistics of each of the dependent and independent variables are given in Table 2 below. The average number of returns is 0.05, and of bad debts is 0.03. Since the book is in Spanish (please see footnote 4), the previous purchases of Spanish language (termed "Hispanic") items becomes a relevant predictor of response. Note that 'Hispanic order number', indicating the number of Hispanic book orders, is comparatively lower (mean of 2.24) compared to total order number (mean of 22.93). The number of payments for related category (Hispanic) products is 1.84, showing that there is some bad debt in historical related category purchases. The average number of previous mailings is 0.36, while the length of residence is 9.86 years on average.

We first present model comparison results, with Model 1 estimating only Eq. 1 and 2, where coefficients are estimated using SUR without historical bad debt information in independent variables. In Model 2, we add one-off bad debt ratio to model 1 to estimate the system of equations. Finally in Model 3 we add Eq. 3 also to the model and estimate the simultaneous system of equations. In Table 3, we present the model comparison which shows the predictive power of using bad debt information. This is especially apparent in the comparison of Models 1 and 2, both of which predict the number of orders and the number of returns. Model 2 however, uses bad debt information (one-off bad debt ratio) to predict the number of returns. Model 2 is clearly better than Model 1 in terms of log likelihood and AIC and BIC measures. Model 3 demonstrates the importance of predicting the number of bad debts in addition to the number of orders and number of returns.

The results are fairly consistent across the three models, so we describe the results of Model 3. In the first panel of Table 4 below we present the results of predicting the number of orders using model 3. The total number of historical orders positively

Table 2 Descriptive statistics—Continuous response model

Variable name	Variables description	Mean	Std	Min	Max
Order	One-off order number	1.1	1.74	0	40
Return	One-off return number	0.05	0.29	0	10
Bad debt	One-off bad debt number	0.03	0.24	0	10
TON	Total order number	22.93	46.58	1	877
$(TON^2)^{1/3}$	Cube root transformed squared total order number	6.38	7.62	1	91.62
HON	Hispanic order number	2.24	3.01	0	46
$(HON^2)^{1/3}$	Cube root transformed squared Hispanic order number	1.44	1.32	0	12.84
ORN	One-off return number of the last 5 transactions	0.23	0.66	0	5
$(ORN^2)^{1/3}$	Cube root transformed squared one time return number of the last 5 transactions	0.19	0.5	0	2.92
TBN	One-off bad debt ratio	0.07	0.25	0	1
$(TBN^2)^{1/3}$	Cube root transformed squared one-off bad debt ratio	0.08	0.26	0	1
Z_{QR}	Number of previous mailings	0.36	0.64	0	5
	Length of residence	9.86	10.23	0	52
Z_{RE}	Number of previous mailings	0.36	0.64	0	5
	Length of residence	9.86	10.23	0	52
	Number of payments for Hispanic products	1.84	2.72	0	45
	Number of payments	2.78	1.6	0	5
Z_{ED}	Number of previous mailings	0.36	0.64	0	5
	Length of residence	9.86	10.23	0	52
	Number of payments for Hispanic products	1.84	2.72	0	45
	Number of payments	2.78	1.6	0	5
	Days since last order	397.31	377.3	0	1826

Table 3 Model comparison—Continuous response

	Model 1	Model 2	Model 3
Order equation	Yes	Yes	Yes
Return equation	Yes	Yes	Yes
Bad debt equation	No	No	Yes
bad debt predictor in Return equation	No	Yes	Yes
Log Likelihood	–1298.69	–1174.00	9807.80
Number of parameters	23	27	44
Number of observations	10,023	10,023	10,023
AIC	2643.38	2402.00	–19,527.60
BIC	2809.27	2596.74	–19,210.24

impacts the number of orders, as expected. The number of orders of Hispanic items negatively impacts the number of orders when the non-linearity is taken into account. The positive significant coefficient for the number of previous mailings suggests that the firm has been targeting customers who are of higher potential. Both the number of recent returns (in the last 5 transactions) and the ratio of bad debts have positive effect on the number of orders. This shows the importance of accounting for bad debt information in predicting the number of orders, since those with a tendency to bad debt tend to place a larger number of orders.

In the second panel we present results of predicting the number of returns. The number of recent returns (in the last 5 transactions) is the strongest predictor of the number of returns. The number of previous orders, especially previous orders of Hispanic products, also is predictive of the number of returns. However, the interesting result from this study's point of view is that *there is a positive relationship between the ratio of bad debts and the number of returns*. In other words, customers who have more bad-debts tend to return a lot. This is different from the firm's belief that customers who return are conscientious while those who bad debt are not. In the third panel we present results of predicting the number of bad debts. The main predictors of the number of bad debts are historical behaviors of bad debt (one-off bad debt ratio) and payments, especially for Hispanic products. The number of orders, especially that of Hispanic products, is also positively related to the number of bad debts.

It has been difficult for the firm to use characteristics of a "returner" to model for returns, because they returned products due to the gap between product quality and their expectations. It may help the firm to predict returns with customers' bad debt information as we find evidence for a positive relationship between the return and bad debt options. Our finding raises the question involving measurement of this effect: how will incorporating information about the interaction of bad debt write-offs and returns impact targeting strategies of a direct marketing firm? To answer this important question, we next utilize a discrete response approach that will estimate probabilities of response of individual customers under different information

Table 4 Demand results—Continuous response

Number of Orders Equation	DV = One-off order number			Number of Returns Equation	DV = One-off return number			Number of Bad Debts Equation	DV = One-off bad debt number		
Independent variables	**Coefficient**	***t* value**	**Net effect**	**Independent variables**	**Coefficient**	***t* value**	**Net effect**	**Independent variables**	**Coefficient**	***t* value**	**Net effect**
Intercept	**0.537**	13.407		Intercept	**0.024**	3.459		Intercept	**0.016**	2.630	
Total order number	**–0.006**	–4.702	0.619	One-off recent return number	**–0.245**	–10.226	0.039	One-off bad debt ratio	**–3.096**	–22.605	0.030
Total order number.^(2/3)	**0.080**	9.592		One-off recent return number.^(2/3)	**0.505**	16.091		One-off bad debt ratio.^(2/3)	**3.409**	25.210	
Hispanic order number	**0.309**	16.776	–0.110	Total order number	0.000	–0.576		One-off recent return number	–0.032	–1.735	
Hispanic order number.^(2/3)	**–0.558**	–13.235		Total order number.^(2/3)	**0.003**	2.268		One-off recent return number.^(2/3)	0.035	1.410	
One-off recent return number	**–1.027**	–7.125	0.056	Hispanic order number	**0.032**	7.018	0.015	Total order number	**–0.001**	–4.210	0.018
One-off recent return number.^(2/3)	**1.544**	8.140		Hispanic order number.^(2/3)	**–0.039**	–5.427		Total order number.^(2/3)	**0.006**	4.812	
Previous mailings	**0.207**	8.178		Previous mailings	0.006	1.503		Hispanic order number	**0.018**	5.070	0.012
Length of residence	**0.011**	7.109		Length of residence	0.000	1.286		Hispanic order number.^(2/3)	**–0.020**	–3.491	

(continued)

Table 4 (continued)

Number of Orders Equation	DV = One-off order number			Number of Returns Equation	DV = One-off return number			Number of Bad Debts Equation	DV = One-off bad debt number		
Independent variables	**Coefficient**	***t* value**	**Net effect**	**Independent variables**	**Coefficient**	***t* value**	**Net effect**	**Independent variables**	**Coefficient**	***t* value**	**Net effect**
One-off bad debt ratio	**–10.807**	–10.203	0.066	Hispanic payment number	**–0.012**	–3.721		Previous mailings	0.003	0.896	
One-off bad debt ratio.^(2/3)	**11.382**	10.898		Payments	**–0.011**	–5.373		Length of residence	0.000	1.722	
				One-off bad debt ratio	**–1.004**	–5.711	0.003	Hispanic payment number	**–0.009**	–3.284	
				One-off bad debt ratio.^(2/3)	**1.011**	5.826		Payments	**–0.012**	–7.359	
								Days Since Last Order	0.000	1.081	

conditions, and develop targeting strategies based on these estimated individual-level probabilities.

3.2 Discrete Response Model

While the previous section modeled continuous response of customers, targeting algorithms used in direct marketing are often based on customers' discrete response to firm direct mailings, which may take the form of not ordering ('no order'), returning ('return'), ordering but defaulting on payment ('bad debt') and ordering and paying for the product('paid'). A discrete choice modeling framework based on the multinomial logit model is thus suitable for modeling this discrete customers response. Figure 1 shows the three possible structures for the logit model according to the customers' decision making process. We model customer choice among the four options in a single stage as in Fig. 1. As discussed in the Web Appendix, this single stage model fits the data in our empirical context better than the alternative choice hierarchies shown in Figs. 2 and 3. Our discrete choice model incorporates heterogeneity in customer response, and also incorporates the firm's targeting policy in the two models to correct for possible endogeneity.

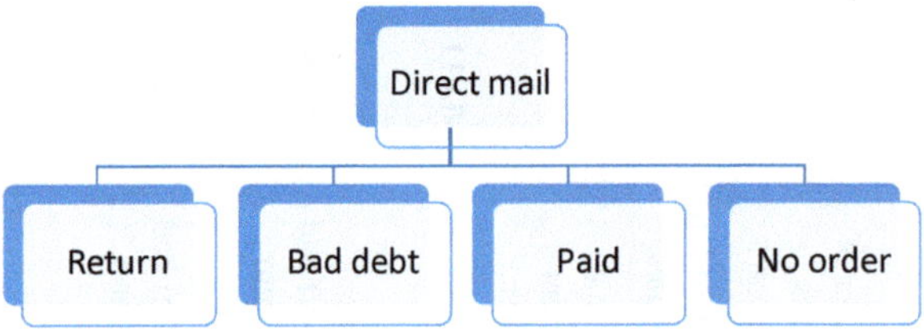

Fig. 1 Model structure 1

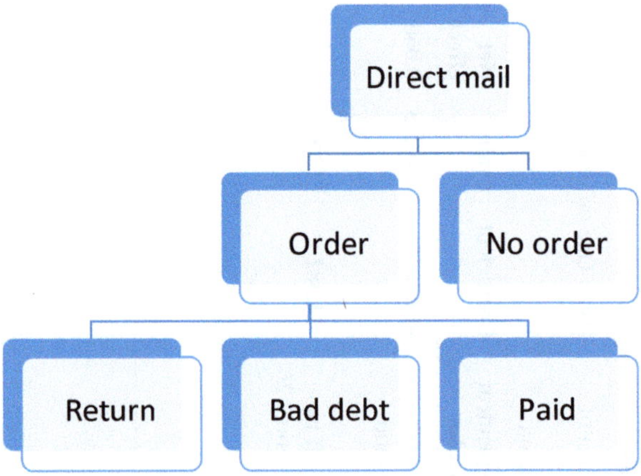

Fig. 2 Model structure 2

Fig. 3 Model structure 3

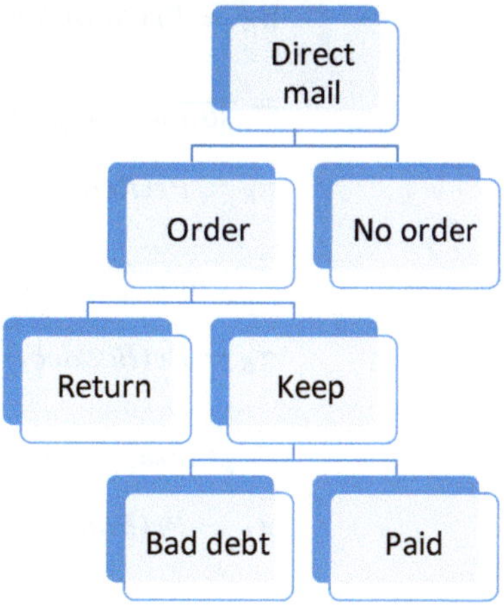

3.2.1 Logit Choice Model

We utilize the logit model as the basic building block for our discrete choice modeling framework. As shown in Fig. 1, there are four discrete choice options (no order, return, bad debt, and paid). As is standard, the intercept is normalized to zero for the last alternative (no order choice) for identification. Intercepts for the other three alternatives (return, bad debt and paid choice) are common across customers and capture the unobserved cost of each alternative that must be overcome for the relevant choice to occur. Consider customer *i* who is mailed the promotion t_i times, t_i=1,..., T_i, where T_i could be 1,2,3,4,5,6. We assume the utilities of the four response alternatives for customer*i* who receives mailings *t* times as:

$$U(Return)_{it} = R + \mu_{R_{it}} + \varepsilon_{R_{it}} \tag{4}$$

$$U(Baddebt)_{it} = B + \mu_{B_{it}} + \varepsilon_{B_{it}} \tag{5}$$

$$U(Paid)_{it} = P + \mu_{P_{it}} + \varepsilon_{P_{it}} \tag{6}$$

$$U(Noorder)_{it} = 0 \tag{7}$$

R, B, and P are the intercepts for the three options, while $\mu_{R_{it}}$, $\mu_{B_{it}}$, and $\mu_{P_{it}}$ capture the influence of the independent variables on each option. The independent variables are listed in Table 2. $\varepsilon_{R_{it}}$, $\varepsilon_{B_{it}}$, and $\varepsilon_{P_{it}}$ are the error terms which are assumed to follow a type $= 1\backslash *$ ROMAN extreme value distribution. The respective probabilities of each of the four choice alternatives for customer i who receives mailings t times is

$$\pi_N = Pr(Noorder)_{it} = \frac{e^{U(Noorder)_{it}}}{e^{U(Paid)_{it}} + e^{U(Noorder)_{it}} + e^{U(Return)_{it}} + e^{U(Baddebt)_{it}}} \tag{8}$$

$$\pi_R = Pr(Return)_{it} = \frac{e^{U(Return)_{it}}}{e^{U(Paid)_{it}} + e^{U(Noorder)_{it}} + e^{U(Return)_{it}} + e^{U(Baddebt)_{it}}} \tag{9}$$

$$\pi_B = Pr(Baddebt)_{it} = \frac{e^{U(Baddebt)_{it}}}{e^{U(Paid)_{it}} + e^{U(Noorder)_{it}} + e^{U(Return)_{it}} + e^{U(Baddebt)_{it}}} \tag{10}$$

$$\pi_P = Pr(Paid)_{it} = \frac{e^{U(Paid)_{it}}}{e^{U(Paid)_{it}} + e^{U(Noorder)_{it}} + e^{U(Return)_{it}} + e^{U(Baddebt)_{it}}} \tag{11}$$

Next we set up the likelihood function that will be maximized to estimate the model coefficients. The indicator variables capturing the four mutually exclusive customer choices will be as follows. Y_P is the customer's choice of paying for the product, taking a value $Y_P = 1$ if paid, otherwise $Y_P = 0$. Y_N is the customer's choice of not ordering to the product: $Y_N = 1$ if no order, otherwise $Y_N = 0$. Y_R is the customer's choice of returning the product: $Y_R = 1$ if returned, otherwise $Y_R = 0$. Y_B is the customer's choice of bad debt: $Y_B = 1$ if bad debt, otherwise $Y_B = 0$.

$$\ln(\text{Likelihood}) = \ln\begin{pmatrix} (\pi_P)^{Y_P} \times (\pi_N)^{Y_N} \times (\pi_R)^{Y_R} \times (\pi_B)^{Y_B}(1-\pi_P)^{(1-Y_P)} \\ \times(1-\pi_N)^{(1-Y_N)} \times (1-\pi_R)^{(1-Y_R)} \times (1-\pi_B)^{(1-Y_B)} \end{pmatrix} \tag{12}$$

The model is estimated by the maximum likelihood procedure which selects the set of values of the model parameters that maximizes the likelihood function. In this way, we can obtain a logit model that fits best with the observed data. With individual level panel data, it is particularly important to account for possible biases due to heterogeneity and endogeneity, and we describe how our model incorporates these aspects below.

3.2.2 Conceptual Framework and Variables Selection

Since direct response firms use discrete choice models to develop their targeting algorithms, we develop a conceptual framework in Fig. 4 to describe the drivers of such targeting policies. While the variables will be developed using this framework for the discrete choice models, the framework in Fig. 4 is also applicable to the continuous response models developed in the previous sub-section.

The no order option is set as the baseline, and we select four to five independent variables for each of the other three options (return, bad debt, and paid), as listed

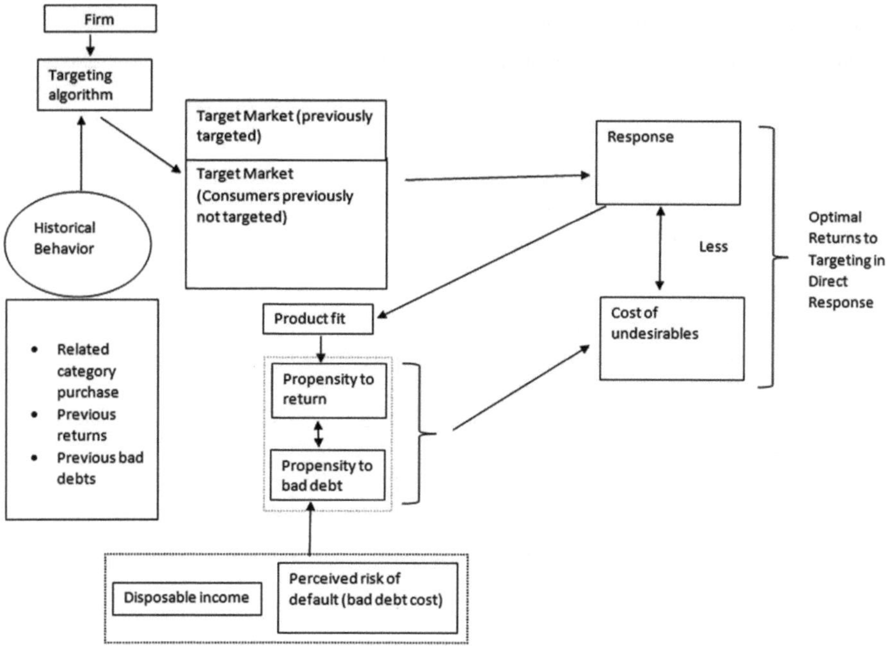

Fig. 4 Direct response targeting tradeoffs and drivers

in Table 5. The independent variables are classified into two types. The first type of variable shows the information of the firm's target activities (number of previous mailings). The second type of independent variables include customers' historical behavior of previous choices (return, bad debt, and paid) which will accommodate lagged interactions between the different choices. For example, a customer's past return behaviors could influence his current bad debt choices. These variables are selected to be consistent with the targeting framework shown in Fig. 4. The details of the independent variables for each of the three options (return, bad debt, and paid) are discussed below.

There are four variables for the return option: previous mailings, bad debt number, Hispanic payment number, and one-off return amount. Previous research (van Diepen, Donkers & Franses, 2009a) shows that the extra mailings would irritate customers, leading to a lower response rate. So here we expect the number of previous mailings the customer received in the past to have a negative influence on his return option. Hispanic payment number also has a negative influence on a customer's return option in the continuous response study. Since the customer who frequently returned products in the past is more likely to return products, the one-off return amount is expected to positively influence a customer's probability of return. The one-off bad debt number is a variable indicating the customer's historical bad debt behavior. This variable is expected to positively affect probability of return.

Table 5 Descriptive statistics—Discrete response model

Response Model	Variable	Description	Mean	Std	Min	Max
Return	Previous mailings	The number of mailings the customer received in the past	0.36	0.64	0	5
	One-off write-off number	The number of one-time write-offs in the past	0.03	0.24	0	10
	Hispanic payment number	The number of payments for Hispanic products	1.84	2.72	0	45
	One-off return amount	The dollar amount of one-time returns	1.51	8.45	0	285.69
Bad debt	Previous mailings	The number of mailings the customer received in the past	0.36	0.64	0	5
	Payments	The number of payments in the last 5 transactions	2.78	1.60	0	5
	One-off write-off ratio	The ratio of one-time write-off count to total write-off count	0.07	0.25	0	1
	One-off return number	The number of cancels or returns in the last 5 one-shot transactions	0.26	0.70	0	5
Paid	Previous mailings	The number of mailings the customer received in the past	0.36	0.64	0	5
	Book write-off number	The number of write-offs for books	0.22	1.03	0	25
	Hispanic order number	The number of orders for Hispanic products	2.24	3.01	0	46
	Total return number	The number of all of the returns in the past	1.74	5.05	0	110

Four variables are selected as drivers of bad debt: previous mailings, payments, one-off bad debt ratio, and one-off return number. As mentioned above, we expect previous mailings to negatively influence a customer's probability of response. The one-off bad debt ratio is expected to positive influence probability of bad debt, while the payments variable is expected to have negatively influence probability of bad debt. These are based on arguments of inertia in customer behavior, i.e. customers who exhibited bad debt in the past continue to exhibit bad debt behavior, while those who tend to pay in the past tend to continue paying for the product in future. Note that this may just represent observed heterogeneity in these behaviors over time rather than state dependence. The one-off return number is included here to explore the relationship between the bad debt and return, and we expect this relationship to be positive, since there is a positive link between these two undesirable customer behaviors.

Four independent variables are used to parameterize the utility for the paid option: number of previous mailings, book bad debt number, Hispanic order number, total return number. The number of previous mailings variable is expected to have a negative influence on the paid option (van Diepen, Donkers & Franses, 2009a). The book bad debt number and total return number are variables concerned with the customers' historical bad debt and return behavior. We expect the customers who frequently returned or wrote off in the past to be less likely to pay for the product. Next we describe the discrete choice model components.

3.2.3 Heterogeneity

We incorporate heterogeneity in the intercepts R, B, P by assuming that they follow a multivariate normal distribution across customers:

$$\beta_1 = \begin{bmatrix} R \\ B \\ P \end{bmatrix} \sim N(\mu_1, \Omega_1) \tag{13}$$

This model has been termed the 'mixed logit' model by Train (2009), and simulated maximum likelihood method can be used to estimate the mixed logit model. Here the expected value of the likelihood is replaced by an arithmetic mean calculated from the sample. If D draws (in our application D equals 20) are made from from $N(\mu_1, \Omega_1)$, probabilities can be computed conditional on each draw, and the average yields the simulated probability

$$\breve{P}_{ij} = \frac{1}{D} \sum\nolimits_{r=1}^{D} L_{ij}(\beta^r) \tag{14}$$

3.2.4 Endogeneity

While the model developed so far will capture heterogeneity in customer response behaviors, it does not account for the type of endogeneity problem that has been labeled 'non-random response' modeling (Donkers et al., 2006; Manchanda et al., 2004), which requires the model to account for the firm's targeting strategy. Rather than inefficiently sending out mailings randomly to customers (an assumption of the standard discrete choice model developed above), the firm targets some customers with its mailings by selecting customers based on past observed behaviors. Not accounting for the firm's targeting decisions will result in an upward bias in the effect of a variable on the customer choice variable such as paid choice, bad debt or return. Therefore it is important that the model be augmented to correct for the endogeneity problem caused by the firm's targeting.

Figure 5 shows the number of customers targeted in each of the six waves of mailings in our data. The numbers of customers who received mailings in each wave are 2439, 1901, 1686, 1259, 1084, and 1413. Note that the last wave shows a heightened level of targeting activity on the part of the firm, presumably to try and maximize response. In the first wave, we do not know the firm's full targeting decisions because there is no information about the customers who are not targeted. Consequently we only model customers' responses for the first wave. For the second to sixth wave, we model both customerss' responses and the firm's targeting. The number of customers who were targeted by the firm from the second to sixth wave are: 341, 577, 553, 575, and 736 (shown in Fig. 5). For example, in the third wave, the firm targeted 577 customers from the customers who received mailings in wave one or wave two.

As described earlier we utilize the conceptual framework in Fig. 7 to develop predictors for firm targeting policies; the variables were decided upon based on inputs of the firm tasked with developing the targeting algorithms. We use five independent

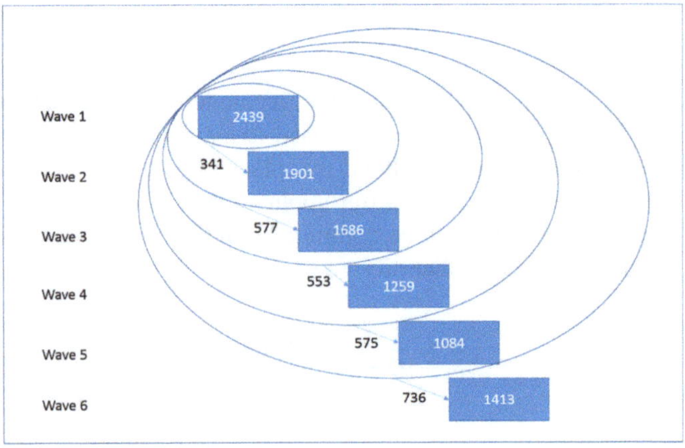

Fig. 5 The numbers of customers targeted in each wave

variables to account for the firm's targeting: *Previous mailings, total return number, recent bad debts, payments*, and whether the customer responded recently (*responded recently*). *Previous mailings* is motivated by Van Diepen et al., 2009b, who found that customers may be irritated by extra mailings, in which case the firm would do well to avoid further targeting of such customers. We therefore expect p*revious mailings* to have a negative effect on the firm's targeting decisions. The remaining variables are related to customers' response behaviors to direct mail. The coefficient signs we expect for these variables are fairly intuitive: *total return number* and *recent bad debts* are expected to negatively influence the firm's targeting, while *payments* and *recently responded* are expected to positively influence the firm's targeting decisions, since the firm prefers to target customers who have a track record of payments.

We use a logit model to describe the firm's targeting. Let I_{it} (t = 2, 3, 4, 5, 6) denote the targeting decisions of the firm for the wave *t*. The indicator variable I_{it} equals 1 if the firm sends a mailing to customer *i* at wave *t*, and equals 0 otherwise. The utility for the firm to send a mailing to customer *i* at wave *t* is:

$$U(I_{it} = 1) = b + \mu_{it} + \varepsilon \tag{15}$$

where

$$\begin{aligned}\mu_{it} = {} & \gamma_1 \text{Previous mailings} + \gamma_1 \text{Total return number} + \gamma_2 \text{Recent bad debts} + \\ & \gamma_3 \text{Payments} + \gamma_4 \text{Recently responded}\end{aligned} \tag{16}$$

The intercept *b* is assumed to follow a normal distribution:

$$b \sim N(\mu_b, \sigma_b) \tag{17}$$

The utility of not targeting a customer is zero.

$$U(I_{it} = 0) = 0 \tag{18}$$

The probability that the firm targets customer i at wave t is

$$\pi_1 = Pr(I = 1)_{it} = \frac{{}_eU(I_{it} = 1)}{1 + {}_eU(I_{it} = 1)} \tag{19}$$

The probability that the firm will not target customer i at wave t is

$$\pi_2 = Pr(I = 0)_{it} = \frac{1}{1 + {}_eU(I_{it} = 1)} \tag{20}$$

The likelihood for the firm's targeting model is

$$(\text{Likelihood})_T = (\pi_1)^I \times (1 - \pi_1)^{(1-I)} \tag{21}$$

Equation 12 provides the likelihood for the customer response model. As in Donkers et al (2006), we can combine the likelihoods for the customer response model from Eq. 12 with the likelihood for the firm targeting model in Eq. 21 to obtain the log likelihood of the full model:

$$\ln(\text{Likelihood})_{Full} = \ln(\text{Likelihood}) + \ln(\text{Likelihood})_T \tag{22}$$

4 Discrete Choice Model Results

The results for mixed logit model which incorporates both heterogeneity and endogeneity correction are shown in Table 6. The results are fairly intuitive, with variables capturing tendency to previously engage in a behavior positively influencing probability of the relevant customer discrete response. For example, one-off return amount (coefficient 0.03; *t*-value 4.999) positively impacts probability of return, one-off bad debt ratio (coefficient 2.003; *t*-value 6.258) positively impacts probability of bad debt. Similarly, Payments (coefficient –0.401; *t*-value –3.677) negatively impacts probability of bad debt while Hispanic order number captures the tendency to buy (and pay for) related product categories.

Apart from these intuitive effects, interactions between the different choices provide some interesting and less obvious predictors of customer discrete response. One-off bad debt number (coefficient 0.459; *t*-value 2.37) positively impacts probability of returns, indicating that customers who tend to bad debt in the past also tend to return the product. This could be due to a malicious intent of procuring the product with the intention of using it and returning it once their needs are satisfied, or due to a genuine lack of fit between the actual product (once it is delivered) and their needs. It is beyond the scope of the data at hand to determine which of these motivations drives their behavior. However, the data shows evidence of this link between historical bad debt and tendency to return. Another less obvious predictive phenomenon is the impact of total return number on the probability of paid choice. While again the exact reason for this predictive link is beyond the scope of the dataset we utilize, this link is not inconsistent with the findings of the recent literature on returns. Kumar and Petersen (2009) discuss the dual impact of returns on customer value, with moderate levels of returns acting as an enabler of relationship building between the firm and the customer and the consequent creation of switching costs for the customer. This beneficial effect of returns up to a threshold is predicated on the customer being well-intentioned and 'genuine' in terms of their return behavior, in the sense that with respect to the purchase, a priori they try to maximize their utility, but a posteriori find that there is lack of fit between the actual product and their needs. This is qualitatively different from the a priori malicious intention of the customer who is either irresponsible in terms of purchase and therefore returns inordinate quantities of products, or who wants to use the product for a particular (social or consumption) occasion and then return the product, effectively gaining 'something for nothing'.

Table 6 demand results—Discrete choice model with heterogeneity and endogeneity correction (*** denotes statistical significance at the 1% level ($p < 0.01$)

Customer response model		
Variable/Model Component	Coefficient	*t*-value
Return choice model component		
Intercept_return	–5.916***	–31.094
Previous mailings	0.246	0.952
One-off bad debt number	0.459***	2.37
Hispanic payment number	0.043	1.068
One-off return amount	0.030***	4.999
Bad debt Choice model component		
Intercept_bad debt	–4.978***	–18.016
Previous mailings	–1.146*	–1.798
Payments	–0.401***	–3.677
One-off bad debt ratio	2.003***	6.258
One-off return number	–0.608	–1.525
Paid Choice model component		
Intercept_paid	–3.883***	–53.575
Previous mailings	–0.11	–0.882
Book bad debt number	–0.051	–0.854
Hispanic order number	0.070***	5.791
Total return number	0.014*	1.728
Heterogeneity covariance (Choleski elements)	0.021	0.14
	0.014	0.09
	–0.018	–0.117
	0.009	0.152
	–0.004	–0.08
	0.007	0.12
Firm Targeting Model		
	Coefficient	*t*-value
Intercept_target	–2.905***	–60.948
Previous mailings	0.563***	11.564
Total return number	–0.007	–1.485
Bad debts in last five transactions	–0.042	–1.301
Payments	0.057***	4.182
Responded recently	–0.097*	–1.875
std variance (heterogeneity)	–0.008	–0.352

There is some evidence from our results that in our empirical context customers may be exhibiting this type of beneficial relationship between returns and paid choice. One other piece of evidence that is suggestive in this context is the coefficient of *one-off return number* in predicting the probability of bad debt choice probability. While this coefficient (–0.608; *t*-value –1.525) is not statistically significant, the negative sign appears to provide some evidence of the possible beneficial impact of returns in this dataset.

The firm targeting model coefficients in the bottom panel of Table 6 illustrate the preoccupations and challenges of a direct marketing firm in terms of its targeting options. The firm heavily mails those who have been mailed earlier (coefficient 0.563; *t*-value 11.564) and those who have made their payments historically (coefficient 0.057; *t*-value 4.182). They also avoid irritating customers by not targeting those who have responded recently (coefficient –0.097; *t*-value –1.875). The fact that total return number and bad debt in last five transactions have negative coefficients approaching significance shows that while the firm is trying to avoid such undesirable customers, it is still casting a wide targeting net as long as the customer has made a sufficient number of historical payments.

5 Discussion

In this section we discuss implications of our demand analyses by conducting simulations utilizing the estimates from our discrete response model.

5.1 Comparative Statics

From Table 6 we can infer that there is a positive relationship between propensity to return and to bad debt, since customers who frequently took on bad debts in the past are more likely to return products. Strategically this raises the interesting question: what would be the impact on firm profits if it does not allow product returns, thus effectively removing the return choice from the customer's choice set? As a first step towards understanding the answer to this question, we simulate probability changes for the three options (bad debt, paid, and no order) after removing the return option, then study the firm's optimal targeting scheme when the return option is removed.

One empirical issue that the researcher needs to deal with in this dataset is the existence of many zeros in the independent variables. For each of the variables, we choose the top 100 customers and the bottom 100 customers after removing the zeroes in the data. For example, in the case of one-off return amount, we sort the 10,023 observations by the one-off return amount variable. The top 100 customers have a mean of 60.381 dollars of one-off return amount, while the bottom 100 customers have a mean of 27.400 of one-off return amount. We calculate the probabilities of the three options (no response, bad debt, paid) for each individual after removing the return option, and list the mean probabilities for the top 100 customers and the bottom 100 customers in Table 7. We also perform t-tests to examine whether the distributions for the top group and the bottom group are different, and p-values are listed in the table.

Our key result from this simulation is that the firm can make greater profits by removing the return option. We first study the three variables concerned with the historical return behavior. The variable one-off return number decreases the probability change for bad debt while increasing the probability change for paid. Since there is no nesting structure in the discrete choice model, customers can consider the return option with the other options simultaneously, which results in the return variables having greater influence. The one-off return number slightly decreases the probability change for the no response option while slightly increasing the probability change for the paid option. Customers who returned a lot in the past are more likely to choose the paid option if the return option is removed, and are less likely to choose the no response option. The number of previous mailings increases the probability change for the no response option and decrease the probability changes of bad debt and paid choice.

Next we study the impact of historical bad debt behavior. The book bad debt number slightly increases the probability of bad debt if the return option is removed. This is not surprising because the customers who frequently took on bad debt can still choose bad debt after the return option is removed. Finally, we look at the variables concerned with historical payment or order behaviors. All these variables (Hispanic payment number, Hispanic order number, and payments) negatively influence the no response probability while positively influencing the paid choice probability. We then rank by the variable one-off bad debt number, and we find that the customers with higher one-off bad debt number are more likely to take on bad debts.

5.2 *Impact on Profits*

Next we simulate the firm's targeting strategy by assuming the costs/profit of the four options and calculate the expected profit of each customer. The customers are targeted on the basis of positive expected profits. There is a mailing cost and targeting cost for no response, and we assume this cost to be $c = \$0.5$ based on current mailing costs. For the bad debt option, there is a cost for the product (a Spanish magazine) in addition to the mailing cost, so we assume the total cost to be $b = \$5$. The return option is even more costly due to handling and restocking cost, so we assume it to be $r = \$8$. It is to be noted that these costs, though seemingly somewhat counter-intuitive, were obtained based on feedback from the management of the direct response company and reflect actual costs for this particular company. The paid option is the only profitable option, and we assume the profit to be $v = \$15$ for each magazine. The number of customers targeted and the profits are listed in Table 8. We find that the total profit is \$7883 if product returns are allowed, but it increases to \$10,628 when product returns are not allowed.

Table 7 Comparative statics—Discrete choice model

Variables		mean	no order	bad debt	paid
One-off return amount	top	60.381	0.892	0.009	0.099
	bottom	27.400	0.934	0.008	0.058
	difference	32.981	–0.042	0.001	0.041
	p-value	0.000	0.003	0.571	0.003
One-off return number	top	3.870	0.911	0.010	0.080
	bottom	1.000	0.937	0.009	0.054
	difference	2.870	–0.027	0.001	0.026
	p-value	0.000	0.002	0.757	0.002
Total return number	top	38.420	0.905	0.010	0.085
	bottom	1.000	0.933	0.013	0.053
	difference	37.420	–0.028	–0.003	0.031
	p-value	0.000	0.000	0.385	0.000
Previous mailings	top	3.120	0.956	0.001	0.043
	bottom	1.000	0.944	0.004	0.052
	difference	2.120	0.012	–0.003	–0.009
	p-value	0.000	0.008	0.000	0.046
One-off bad debt number	top	1.650	0.885	0.040	0.075
	bottom	1.000	0.901	0.037	0.062
	difference	0.650	–0.015	0.002	0.013
	p-value	0.000	0.213	0.730	0.212
One-off bad debt ratio	top	1.000	0.853	0.053	0.094
	bottom	0.361	0.925	0.015	0.061
	difference	0.639	–0.072	0.038	0.033
	p-value	0.000	0.000	0.000	0.013
Book bad debt number	top	7.830	0.903	0.043	0.055
	bottom	1.000	0.919	0.036	0.046
	difference	6.830	–0.016	0.007	0.009
	p-value	0.000	0.123	0.404	0.129
Hispanic payment number	top	18.860	0.868	0.008	0.124
	bottom	1.000	0.927	0.013	0.060
	difference	17.860	–0.058	–0.005	0.064
	p-value	0.000	0.000	0.006	0.000
Hispanic order number	top	20.550	0.856	0.010	0.134
	bottom	1.000	0.925	0.016	0.059
	difference	19.550	–0.069	–0.007	0.075
	p-value	0.000	0.000	0.134	0.000
Payments	top	5.000	0.749	0.004	0.247

(continued)

Table 7 (continued)

Variables		mean	no order	bad debt	paid
	bottom	1.000	0.906	0.020	0.074
	difference	4.000	–0.157	–0.016	0.173
	p-value	0.000	0.000	0.001	0.000

Table 8 Impact of product returns option availability on profits

Scenario	No order	Return	Bad debt	Paid
Product returns allowed	0.851	0.027	0.020	0.101
Product returns not allowed	0.872	N/A	0.021	0.107

Assume the utilities of customers' options are u_1, u_2, u_3, and u_4, then the expected profit increase when product returns are not allowed is

$$\Delta EP = \frac{e^{u_2}}{(e^{u_1} + e^{u_2} + e^{u_3} + e^{u_4})(e^{u_1} + e^{u_3} + e^{u_4})} \times r \times \left(-\frac{c}{r} \times e^{u_1} - \frac{b}{r} \times e^{u_3} + \frac{v}{r} \times e^{u_4} + e^{u_1} + e^{u_3} + e^{u_4}\right) \quad (23)$$

Since $\frac{e^{u_2}}{(e^{u_1}+e^{u_2}+e^{u_3}+e^{u_4})(e^{u_1}+e^{u_3}+e^{u_4})} > 0$ and r > 0, the expected profit will decrease only if $\left(-\frac{c}{r} \times e^{u_1} - \frac{b}{r} \times e^{u_3} + \frac{v}{r} \times e^{u_4} + e^{u_1} + e^{u_3} + e^{u_4}\right) < 0$.

We have found that the firm can increase profit by not allowing product returns with the option cost set at the set of values given by {c = 0.5, b = 5, r = 8, v = 15}. Next, we will change the option cost set to study whether the profit will decrease when product returns are not allowed. We will fix r and change c, b, and v respectively to see how profit changes.

Firstly, we study how profit will change with variation in bad debt cost b. When we fix profit v, we have results shown in Fig. 6. The horizontal axis shows the ratio of bad debt cost to return cost(b/r), while the vertical axis shows the ratio of the profit when product returns are not allowed to the profit when product returns are allowed. We find that the profit ratio slightly decreases when bad debt cost increases. However, whether the profit ratio will be less than 1 is mainly decided by the ratio of targeting cost to return cost (c/r) instead of the ratio of bad debt cost to return cost (b/r). When we fix c, we have results shown in Fig. 7, which are similar to those in Fig. 6. Therefore the bad debt cost has little influence on the profit ratio.

Next we study how profit changes with changes in the targeting (and mailing) cost c. Since b has little effect on profit, we fix b = c. Figure 8 shows that the profit ratio decreases steeply when targeting cost c increases. That is, the company cannot increase profit by not allowing product returns when targeting cost is higher. A possible reason is that customers are less likely to order products when product returns are not allowed. However, the total cost for the company is higher as targeting

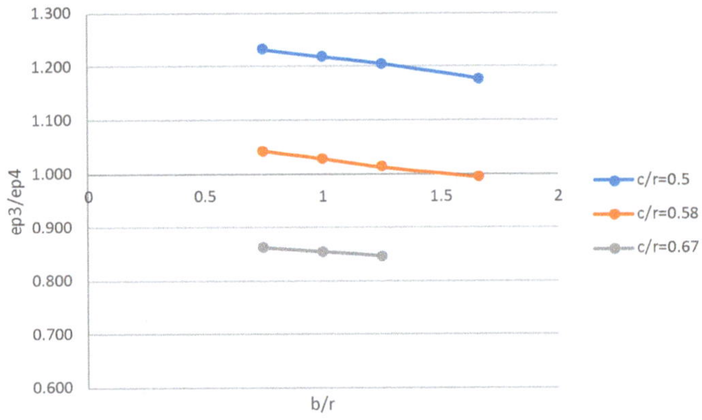

Fig. 6 Profit changes when b changes (v fixed)

Fig. 7 Profit changes when b changes (c fixed)

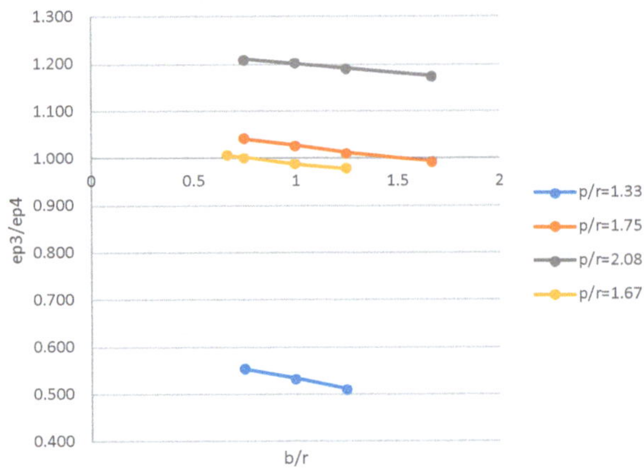

(and mailing) cost increases, and therefore the company can lose profit if product returns are not allowed.

Last we study how profit changes when v changes. Since b has little effect on profit, we fix b = c. Figure 9 shows that when v is high, the company can increase profit by not allowing product returns, which is the main finding of our paper.

Although this analysis has some limitations since it is not a field experiment, the simulation analyses demonstrate that the customers who are likely to return products are likely to pay for the products rather than become bad debts when there is no return option. It should be noted that this result has an intuitive appeal, since customers who take the time and effort to repackage a product, go to the post office and mail it, are more conscientious than customers who bad-debt. This leads us to the conclusion that as long as the firm has a small cost of 'no response' (as captured by mailing and targeting costs), it can increase profit by adopting the strategy of not allowing product returns.

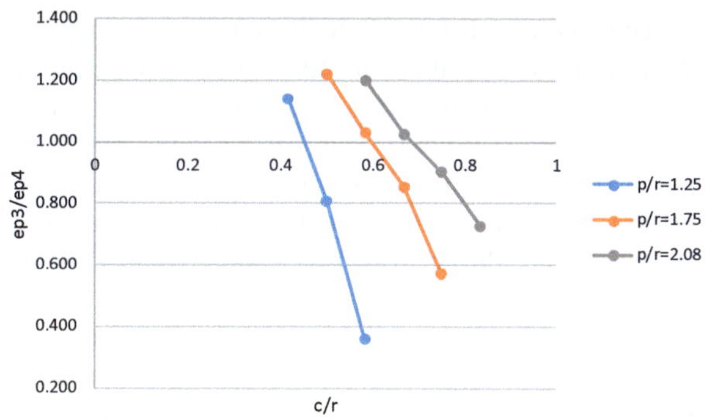

Fig. 8 Profit changes when c changes (b fixed)

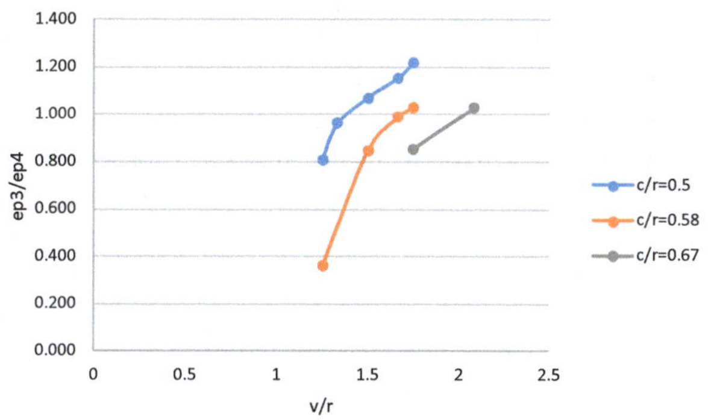

Fig. 9 Profit changes when v changes (b fixed)

5.3 *Return -Bad Debt Tradeoff: Simulation Diagnostics*

The results show that the one-off bad debt number has a significantly positive influence on the return option, while a similar variable one-off bad debt ratio has a significantly positive influence on the bad debt option. Across the continuous and discrete response studies, there is some evidence that historical returns behavior negatively influences propensity to bad debt.

The simulation results show that the customers with high bad debt numbers would still choose the bad debt option if the return option is removed. However, the customers with high return numbers would choose to pay for the products rather than take on bad debts. And the customers with high payment numbers are also more likely to pay for products. Therefore the firm should target customers with more returns and more payments in the past.

The cost of bad debt option seems to have little influence on the profit change when the firm does not allow product returns. Only when the cost of no order is as high as half of the return cost will the firm suffer losses by not allowing product returns. Since the no response cost is generally not very high, the firm can increase profits by not allowing product returns.

There is some limitation to the simulation study. In the simulation study, we obtain the estimates when there are four choices in the model, remove the return option and assume the estimates do not change when the return option is removed. However, as the Lucas' critique (Lucas, 1976) states, economic agents who observe the new policy (no return option is offered) may change their response to the stimulus (targeted offers). In other words, the estimates of the demand model that captures customer response may differ from that estimated in this reduced form model. Therefore, the probability changes for the three options (no order, bad debt, and paid) may be different from those obtained in the demand analysis. There are two methods to address this issue: one, a structural model that builds up customer response from primitives of customers and firm behavior. And two, a field experiment that allocates customers randomly to conditions where the return option is present/absent. We position our work as the first evidence of a positive link between the two undesirable customer behaviors of bad debts and returns, and believe that the structural approach and/or field experiment approach can be fruitful avenues for future research.

Our paper suggests the firm remove the return option, which is at odds with the findings in Anderson, Hansen, Simester (2009). They find that the customers are less likely to respond when the return option is not allowed, thus suggesting the return option should be offered. However, they do not consider the bad debt option, which in our application costs much more than the no order option. Table 7 shows that the targeted customers have a higher mean probability of no order when the return option is removed, which is consistent with the findings in the earlier literature. However, the mean bad debt probability of the targeted customers only increases a little when product returns are not allowed. As a result, the loss caused by the product returns and bad debts are greatly reduced when the return option is not offered. This implies that even though the customers are less likely to respond on average, the firm can make a greater profit by not offering the return option.

5.4 Return-Bad Debt Tradeoff: Discussion

Customers could react in several different ways when they receive a catalog. They could ignore the catalog or they could place an order from it. In the latter case, they may choose to keep the product or return it. Customers could be conscientious, honestly frustrated or exploitative in their return behavior, and the difficulty of managing this process stems largely from the difficulty in ascertaining 'good' intentions from 'bad' ones. Studies on product returns have generally been conducted for target segments for whom payment is not an issue, as most payments are made by credit card, and the decision regarding credit-worthiness of the customer is predetermined by the company that issued the credit card on which the purchase was made. When such merchants who rely mostly on credit cards for payment engage in targeted marketing, their targeting algorithms do not need to consider whether the customer is credit-worthy. In contrast, firms engaging in direct response marketing often target a segment of the market that are of lower income, and who may often not have, or use credit cards for discretionary purchases. In these types of 'bill me later' transactions, the product is purchased first, and the payment is made, usually by check, on receipt of the product. In these direct response transactions, the estimation of credit-worthiness of the customer is endogenized into the targeting algorithm. Motivations for the other undesirable behavior of 'bad debt', where customers just keep the product without paying for it, are less ambiguous, stemming either from sheer forgetfulness or a clear-cut intention to cheat the firm. Yet bad debts can also be difficult to predict and to manage, especially in the case of "buy-now-pay-later" payment option that is often offered for small ticket items. While tracing out motivations for these undesirable behaviors is an important research question, it is beyond the scope of the data utilized in this study. Our analyses do present some findings that throw some light on possible inter-relationships between these undesirable behaviors and raise questions about whether this phenomenon could be unique to customers with limited or no credit card debt availability. Customers could be utilizing separate mental accounts (Thaler, 1985) for bad debt and returns where they trade off actual costs as well as psychic costs they might incur in each scenario. Optimization of such costs across transactions could lead to some of the linkages between undesirable behaviors that our analysis uncovers.

Our results support the findings of the literature on returns by providing a further piece of evidence—that customers who return are largely customers who tend to pay rather than bad debt. This is so even among customers with low (or no) access to credit card debt. Our recommendations that firms should seriously consider dropping the return option for at least some of their 'bill me later' campaigns is based on the economics of return costs rather than on the profile of customers who tend to return products. Our research finding may not apply to bigger ticket items, where the temptation to bad debt may be higher if there is no return option. However, it should be noted that in those cases the economics of the transaction would mean that there are likely to be greater collection efforts on the part of the firm. The other interesting implication of our finding has to do with the online 'bill me later' options such as the one proposed by the company acquired by Paypal in 2008. It is to be noted that in this case the linkage between undesirable behaviors that we detect in the case of a single firm is still likely to exist. However, the tradeoff of the greater risk of bad debt versus higher response rate would now occur across two firms. These are the online 'bill me later' firm that guarantees credit and is responsible for collection of debt, and the targeting firm that enjoys the higher response rate, and that pays some of the returns from this as a fee to the 'bill me later' company.

6 Conclusion

This paper studies the customers' responses to direct mail considering both the bad debt and return options by utilizing the historical transaction information variables that influence the customers' return, bad debt, and paid options. We find that customers with high payment numbers are less likely to make returns or bad debts, while those who frequently took on bad debts in the past are more likely to take on bad debts and more likely to return products. Also, customers who ordered a lot of products in the past are less likely to pay for the products. Overall, we believe that the challenges to targeted marketing uncovered by this study constitute the beginning of a possibly rich set of research questions that can arise from studying the unique challenges and variants of marketing concepts that are applied in marketing to customers with limited or no access to credit card debt.

The paper also studies the impact of removing the product return option. The results show that the customers with high return numbers in the past are more likely to pay rather than take on bad debts if the product return is not allowed. The customers with high write-off numbers may become bad debts, while the customers with high payment numbers are more likely to pay for products. Therefore customers with high return numbers and high payment numbers are the most profitable for the firm. We also demonstrate that the firm can increase its profit by forbidding product returns if the no response cost is not too high.

There are some limitations in this research. First, we only consider the heterogeneity in the intercepts in this paper. We can improve this by assuming all of the coefficients to follow distributions in future research. Second, the customer demand functions estimated can be considered a first approximation due to the reduced form approach and applicability of the Lucas Critique. Nevertheless, this paper provides the first step in the measurement of the tradeoff between bad debts and returns in the targeting literature. In spite of the limitations, the paper contributes to product returns in direct mail by considering the bad debt option together with the return option. We find some positive interactions between the customers' historical return behaviors and their possibilities to take on bad debts. We also provide evidence for a counter-intuitive strategy whereby the company reduces return costs by not allowing the product returns, and thus increase profits. Firms that resort to this option may explore other avenues for relationship building such as loyalty programs to substitute for the relationship-building component of returns behavior described in Petersen and Kumar (2009).

References

Anderson, E. T., Hansen, K., & Simester, D. (2009). The option value of returns: Theory and empirical evidence. *Marketing Science, 28*(3), 405–423.

Ansari, A., Mela, C. F., & Neslin, S. A. (2008). Customer channel migration. *Journal of Marketing Research, 45*(1), 60–76.

Bodapati, A. V. (2008). Recommendation systems with purchase data. *Journal of Marketing Research, 45*(1), 77–93.

Bower, Amanda B., & James G. Maxham III. (2012). Return shipping policies of online retailers: Normative assumptions and the long-term consequences of fee and free returns. *Journal of Marketing, 76*(5), 110–124.

Bruene, J. (2008). *eBay Acquires Bill Me Later for Almost $1 Billion.* http://finovate.com/ebay_acquires_bill_me_later_for_almost_1_billion/. Accessed November 1, 2016.

Donkers, B., Paap, R., Jonker, J. J., & Franses, P. H. (2006). Deriving target selection rules from endogenously selected samples. *Journal of Applied Econometrics, 21*(5), 549–562.

Dong, X., Manchanda, P., & Chintagunta, P. K. (2009). Quantifying the benefits of individual-level targeting in the presence of firm strategic behavior. *Journal of Marketing Research, 46*(2), 207–221.

Durkin, Thomas A. (1970–2000). *Credit cards: Use and consumer attitudes.* https://www.federalreserve.gov/pubs/Bulletin/2000/0900lead.pdf. Accessed Jan 8, 2017.

Gilbride, Timothy J., Jeffrey Inman, J., & Melville Stilley, K. (2015). The role of within-trip dynamics in unplanned versus planned purchase behavior. *Journal of Marketing, 79*(3), 57–73.

Gönül, Füsun F., & Frenkel Ter Hofstede. (2006). How to compute optimal catalog mailing decisions. *Marketing Science,* 25(1), 65–74.

Gümüs, M., Ray, S., & Yin, S. (2013). Returns policies between channel partners for durable products. *Marketing Science, 32*(4), 622–643.

Hartmann, W., Nair, H. S., & Narayanan, S. (2011). Identifying causal marketing mix effects using a regression discontinuity design. *Marketing Science, 30*(6), 1079–1097.

Holmes, T. E (2014). Credit card statistics. http://www.creditcards.com/credit-card-news/ownership-statistics-charts-1276.php. Accessed Jan 27, 2017.

Kumar, A., Salo, J., & Bezawada, R. (2024). The effects of buy now, pay later (BNPL) on customers' online purchase behavior. *Journal of Retailing, 100*(4), 602–617.

Liu, H., Pancras, J., & Houtz, M. (2015). Managing customer acquisition risk using co-operative databases. *Journal of Interactive Marketing, 29*, 39–56.

Lucas Jr, Robert E. (1976). Econometric policy evaluation: A critique. In *Carnegie-Rochester conference series on public policy* (vol. 1, pp. 19–46). North-Holland.

Manchanda, P., Rossi, P. E., & Chintagunta, P. K. (2004). Response modeling with nonrandom marketing-mix variables. *Journal of Marketing Research, 41*(4), 467–478.

Petersen, J. A., & Kumar, V. (2009). Are product returns a necessary evil? Antecedents and consequences. *Journal of Marketing, 73*(3), 35–51.

Rook, D. W. (1987). The buying impulse. *Journal of Consumer Research, 14*(2), 189–199.

Selden, L., & Colvin, G. (2003). *Angel customers & demon customers: Discover which is which and turbo-charge your stock.* Penguin.

Shulman, J. D., Coughlan, A. T., & Canan Savaskan, R. (2010). Optimal reverse channel structure for consumer product returns. *Marketing Science, 29*(6), 1071–1085.

Statista (2013). *Advertising spending of the direct response companies in the United States in 2013, by medium (in thousand U.S. dollars).* https://www.statista.com/statistics/318099/direct-response-ad-spend-medium/. Accessed Jan 27, 2017.

Stavins, J. (2024). *Buy now, pay later: Who uses it and why.* Federal Reserve Bank of Boston Research Paper Series Current Policy Perspectives Paper 2024–3.

Thaler, R. (1985). Mental accounting and consumer choice. *Marketing Science, 4*(3), 199–214.

Train, K. E. (2009). *Discrete choice methods with simulation.* Cambridge University Press.

van Diepen, M., Donkers, B., & Franses, P. H. (2009a). Dynamic and competitive effects of direct mailings: A charitable giving application. *Journal of Marketing Research, 46*(1), 120–133.

van Diepen, M., Donkers, B., & Franses, P. H. (2009b). Does irritation induced by charitable direct mailings reduce donations? *International Journal of Research in Marketing, 26*(3), 180–188.

Vatne, B. (2021). What looming postal increases teach us about the future of omnichannel marketing. https://www.ana.net/blogs/show/id/mm-blog-2021-05-postal-changes-impact-marketers. Accessed January 3, 2025.

Zhang, J. Z., Chang, C.-W., & Neslin, S. A. (2022). How physical stores enhance customer value: The importance of product inspection depth. *Journal of Marketing, 86*(2), 166–185.

Sentiment Analysis of Consumer Reviews Using Enhanced Attention-CNN-BiLSTM Networks

Amandeep Kaur, Amandeep Kaur, and Ayush Deotare

Abstract Extraction of valuable insights from online customer reviews helps in better decision-making process. While sentiment analysis serves this purpose, it faces certain problems like review bias where sentiments are inclined excessively positive or negative leading to reduced accuracy. In this study, we propose a system that utilizes word embeddings in order to capture semantic relationships and integrate a CNN-BiLSTM model to detect sentiment of the review more effectively. Topic-based attention mechanism is incorporated in the proposed system to get more information about the predefined context culminating more precise analysis. Performance analysis of the proposed system is performed using two datasets namely Amazon Reviews dataset and Yelp dataset employing a number of performance metrics. Experimentation results indicate the effectiveness of the proposed system to reduce the effect of review bias and enhance the overall accuracy. This research is advancement in the field of sentiment analysis making use of advanced Deep Learning (DL) techniques, enhancing the understanding of customer sentiment.

Keyword Online reviews · Word embedding · Deep learning · Attention mechanism · Customer sentiment analysis

---

A. Kaur (✉) · A. Deotare
Department of Management Studies, ABV-Indian Institute of Information Technology and Management Gwalior, Gwalior, India
e-mail: adeep5524@gmail.com

A. Deotare
e-mail: ayush.deotare@gmail.com

A. Kaur
Department of Computer Science and Engineering, National Institute of Technology Delhi, Delhi, India
e-mail: amandeepk@nitdelhi.ac.in

B. Basu et al. (eds.), *Data Analytics for Decision Making towards Business Excellence*, Uncertainty and Operations Research, https://doi.org/10.1007/978-981-95-2429-7_6

1 Introduction

In e-commerce sector, online reviews have emerged as a ubiquitous force, wielding immense influence over consumer purchasing decisions. These digital testimonials, often perceived as authentic reflections of consumer experiences, have become an indispensable tool for marketers seeking to gauge consumer perception and optimize product strategies (Chehal et al., 2021; Ullal et al., 2021).

Mining and leveraging these online reviews, business owners can gain an edge in their respective competitive markets by taking rational decisions. In recent years, Natural Language Processing (NLP) seeks researchers' attention to develop solution in these areas by understanding the sentiments of customer textual reviews (Tam et al., 2021).

The term 'Sentiment Analysis' refers to extracting emotions and opinions from the textual data. It has gained widespread recognition not only within the research community but also among various stakeholders, including governments and corporations, for obtaining actionable insights. The sheer volume of online reviews, often viewed as an indicator of product popularity, does not always provide a reliable measure of consumer sentiment (Kim, 2021). The data about the online reviews need careful analysis about various aspects including sentiments to extract meaningful information.

Online reviews have plenty of data that is difficult to handle with traditional models. Machine Learning (ML) and Deep Learning (DL) techniques have shown numerous advantages and are widely used for analysing the online reviews (Huang et al., 2022; Karabila et al., 2023; Lin et al., 2019; Xia et al., 2020). Basically, ML consists of learning models and statistical techniques to perform analysis, classification and prediction tasks. DL includes various multi-layer networks that are capable of handling complex and nuanced language patterns. Recurrent neural network (RNN) processes the time-related information more effectively. Long short-term memory (LSTM) networks and Bidirectional LSTM (Bi-LSTM) have been used successfully in analyzing the sentiments in a variety of problems as these networks are capable of utilizing the information bidirectionally ((Murugaiyan & Uyyala, 2023; Tam et al., 2021)). Integration of attention mechanisms further improves the performance of sentiment analysis tasks by emphasizing on important words (Baziotis et al., 2017; Shin et al., 2016; Wang et al., 2016; Yin et al., 2016). Topic level attention mechanisms emphasize on text that is relevant to the specific topic (Huang et al., 2021).

Online review ratings, often touted as a measure of product quality, are susceptible to subjectivity and extremity bias. Consumers who are most inclined to leave reviews are those with extreme experiences, either exceptionally satisfied or exceedingly dissatisfied. This phenomenon leads to an over-representation of extreme values, skewing the distribution and potentially misrepresenting the sentiments of the broader consumer base (Jiang et al., 2023; Obiedat et al., 2022). So, it is important to remove the biases from the data to draw meaningful inferences about product success. Furthermore, the rise of platforms like Amazon, where users can mark reviews as helpful, adds an additional layer of complexity to the analysis. The number of reviews marked as helpful might offer insights into the collective sentiment of the reviewing community, but it also necessitates a thorough investigation to determine the impact of such metrics on the overall assessment of products.

This research work seeks to bridge the gap between the assumptions made by marketers regarding online reviews and the nuanced reality of consumer behavior and sentiment in the digital age. This research work aims to unravel the complexities surrounding online reviews by using DL models in understanding and predicting customer sentiment. Furthermore, biases in the data are removed before feeding them to predictive models using sampling methods to get unbiased results.

The main contributions of this research are:

1. We develop a novel hybrid model that integrates a topic-based attention mechanism with a CNN-BiLSTM architecture. The topic-attention mechanism is proposed to effectively adjust the weights of text subsequence hidden representations, enabling the model to focus on the most relevant topics in the textual review. This model learns bidirectional dependencies in textual reviews, enabling it to capture high-level semantic and contextual features more effectively.
2. We utilize word embedding techniques to capture nuanced semantic relationships between words. We also apply oversampling technique, SMOTE, Borderline SMOTE and SVMSMOTE methods to address class imbalance in the dataset, ensuring more robust and balanced sentiment classification.
3. We conduct rigorous experimental validation of the proposed model using well-established datasets, specifically the Yelp and Amazon review datasets, to evaluate and demonstrate the model's effectiveness using evaluation metrics.

The remaining paper is structured as follows. Section 2 elaborates the related work. The proposed model is discussed in Sect. 3 and Sect. 4 discusses the experimentation. Section 5 elaborates the results while Sect. 6 gives deployment and real-time inference. Finally, Sect. 7 presents the conclusion and future research directions.

2 Related Works

In the landscape of research on sentiment analysis of online reviews, many notable contributions come from various researchers. García-Moya et al. (Garcıa-Moya & H. Anaya-S´anchez, & R. Berlanga-Llavori, 2013), in 2013, put forth a methodology that made use of customer reviews to identify opinions. Their methodology, emphasizing opinion mining, effectively retrieved product features and associated opinions in multiple languages. Notably, their approach did not require a training set of product features, contributing to the development of feature-based opinion extraction methods. Lei et al. (2016), in 2016, proposed a recommendation system that uses sentiment measurement in addition to other factors. Authors aimed to enhance prediction accuracy in recommender systems by incorporating sentiment-based factors. Authors utilized not only user sentimental information but also product reputation for the analysis. Their approach highlighted the significance of sentiment information in optimizing recommendation algorithms, thus emphasizing the ongoing need for nuanced improvements in the field. Furthermore, Yang et al. (2023), in 2019, developed a sentiment-driven model for online ranking of products and performance evaluation across various platforms. The attributes are mined and sentimental values are computed using ML algorithms. In contrast, Kim (2021), in 2021, presented a framework that made use of online reviews to analyze customer sentiments aiming to address the limitations of traditional metrics in measuring customer sentiment. The study proposed sentiment scores as a more reliable metric and underscored the importance of leveraging online reviews for a more nuanced understanding of customer sentiments.

2.1 ML for Sentiment Analysis

The authors utilized a range of ML algorithms to analyze sentiments. Barbado et al. (2019), in 2019, put forth a framework that analyzed the fake reviews by performing the sentiment analysis. Authors prepared a dataset by taking the fake news data from Yelp in the field of consumer electronics and refining it with a scraping process. User centric and review centric features were considered and the performance of five ML models in fake review detection was compared. Khedkar and Shinde (2020), in 2019, presented a framework to classify the review into two categories namely Praise and Complaint utilizing Linguistic feature-based approach. The presented approach seems helpful for the business owners to satisfy the customer expectations and demands. The authors employed four ML algorithms and evaluated their performances.

Patil et al. (2023), in 2023, presented a system for sentiment classification of twitter data available in Kannada language. Authors carried out comparative analysis using six ML algorithms. After data pre-processing, the features were extracted using the Term frequency-inverse document frequency model before passing the data to the classifier. Li et al. (2023), in 2023, put forth a system to predict restaurant survival using aspect-based sentiment analysis that will guide the business owners to make better decisions. The authors employed a conditional survival forest model to analyze online reviews in a total five categories, analyzed their performance and made better business strategies.

Dervenis et al. (2024), in 2024, explored the use of big data and AI to improve educational outcomes by analyzing student feedback through sentiment analysis. Authors compared ML and lexicon-based methods on performance, multilingual support, context sensitivity, and the impact of a neutral sentiment class. Bisht et al. (2024), in 2024, put forth a study that examined Twitter as a tool for real-time communication, proposing methods to identify trending hashtags and analyze their sentiment dynamics over six months. Kumar et al. (2024), in 2024, presented a study to predict hotel ratings from reviews using NLP to extract features and an ensemble learning approach for enhanced accuracy. It compares seven classifiers and three embedding methods with TF-IDF and Logistic Regression with Cross Validation (LRCV) achieving a high value of accuracy. Rahman et al. (2024), in 2024, proposed a model to classify feedback and comments from various online platforms by proposing an ensemble ML model. By integrating ensemble techniques like stacking and soft-voting, the model achieved superior accuracy.

Additionally, Qureshi et al. (2023), in 2023, presented a study addressing the lack of large datasets by scraping 251,661 app reviews and applying ML models, with SVM achieving the highest accuracy. Authors also highlighted the effectiveness of using cleaned reviews and statistical features, while suggesting opportunities for further research by expanding the dataset and exploring additional features like metadata.

2.2 DL for Sentiment Analysis

DL has potential applications across various fields, including sentiment analysis. This section explores several contributions made by researchers who have leveraged DL techniques in sentiment analysis. For instance, Rahman et al. (2019), in 2019, presented the use of a hybrid model using CNN and LSTM to perform sentiment analysis for movie reviews. After data pre-processing, text data is converted into vectors having numeric data using Word2Vec. CNN-LSTM along with dropout technology and rectified linear unit gives better results by evaluating long-term dependencies. Experimental results corroborate its performance on two datasets of movie reviews. Additionally, Lin et al. (Xia et al., 2020), introduced a novel approach with Bi-LSTM and Multi-Head Attention mechanism to capture nuanced sentiments, demonstrating improved performance.

In 2020, Jelodar et al. (2020) proposed a system to perform sentiment analysis of the online opinions present on COVID-19. The semantic was extracted from the text using topic modeling that employed Latent Dirichlet Allocation (LDA) and then passed to Bi-LSTM model for the classification of sentiments. In 2022, Jain et al. (2022) proposed a novel ensemble LSTM technique-based Airline recommendation system. The system used LSTM and Decision Tree as an ensemble and then Logistic Regression was performed on their outputs to get the final outcome. It includes two modules in which the first module analyzes the sentiments of customer reviews while the other uses service aspect ratings. Huang et al. (2022) introduced the SentiCNN model, a Lexicon-Based Sentiment CNN tailored for online review analysis. Their work successfully integrated sentiment and contextual information, yielding improved performance in sentiment analysis for online reviews. However, while the SentiCNN model showcased advancements, certain limitations persisted in the broader field.

Karabila et al. (2023), in 2023, utilized the BERT model to get a better recommendation system for predicting customer preferences with an enhanced accuracy. Arbane et al. (2023), in 2023, put forth a model that performed sentiment analysis on COVID-19 by analyzing public beliefs and thoughts. Authors used Bi-LSTM to extract sentiment out of the text using two datasets taken from Twitter and Reddit social networking sites. Before forwarding the data to Bi-LSTM, data was splitted into words using tokenizer and then sentiments of data were determined using Bi-LSTM. Furthermore, Anbumani and Selvaraj (2023), put forth an activation function namely sigTan-Beta for CNN to get better results with enhanced accuracy. The proposed system pre-processes the data initially and then converts into a vector employing Word2Vec. Finally, CNN was utilized for the extraction of low and high level of features and classifies them as positive or negative reviews. Table 1 summarizes the literature related to sentiment analysis.

2.3 Attention Mechanisms

As we explore the landscape of neural network architectures, attention mechanisms have enhanced the model's ability to focus on most relevant inputfeatures significantly. Yin et al. (2016), in 2016, proposed the use of an attention-based CNN model for sentence pairs demonstrated improved performance compared to CNNs without attention mechanisms. Shin et al. (2016), in 2016, further explored embedding attention vectors, incorporating sentiment scores of sentiment words, to improve the accuracy of sentiment analysis tasks. However, the study fell short in addressing the potential insufficiency of embedding attention vectors developed for document-level analysis, particularly in utilizing sentiment information effectively. Wang et al. (2016), in 2016, proposed a model that takes different aspects as input and incorporates an attention mechanism in order to emphasize on various segments of a sentence. Baziotis et al. (2017), in 2017, advocated for the application of Bi-LSTM with attention (D-LSTM) to enhance the model performance. Du et al. (2019), in 2019, integrates a RNN with a convolution-based attention model. The field has witnessed the emergence of various novel methods over the past few years. Researchers increasingly recognize the benefits of ML techniques and have successfully employed them to analyze sentiments from text data, thereby enhancing classification performance. Shen et al. (2018), in 2018, proposed the ON-LSTM, a novel LSTM variant tailored for addressing NLP challenges. Notably, the neurons in ON-LSTM are meticulously ordered to express more intricate information.

Despite advancements in sentiment analysis, significant gaps persist, particularly in handling complex patterns, scalability, abstraction of high-level features and interpretability of sentiments. While attention mechanisms Improve the model's capability to concentrate on key aspects, they still struggle to fully capture complex sentiments, which require deeper understanding. Additionally, existing sentiment analysis shows limited generalization in the context of business product reviews. This research work aims to bridge these gaps by integrating advanced neural architectures to significantly enhance performance in sentiment analysis tasks. The proposed approach ensures a deeper contextual understanding and adaptability to specific domains. Additionally, topic-based attention mechanisms improve the transparency

Table 1 Summary of related work

Sr. No	Context	Authors	Methodology used	Dataset
1	Fake Review Detection	Barbado et al. (2019)	Logistic Regression (LR), Decision Tree (DT), Random Forest (RF), GNB (Gaussian Naive Bayes), AB (AdaBoost)	Yelp
2	Praise or Complaint Classification	Khedkar et al. (2020)	RF, K Neighbours Classifier (KNN), SVC (Support Vector Classifier), and Multilayer Perceptron (MLP), a Linguistic feature-based approach	Amazon product reviews dataset
3	Kannada Twitter Sentiment Analysis	Patil et al. (2023)	LR, Stochastic Gradient Descent (SGD), KNN, Multinomial Naive Bayes (MNB), GNB Classifier	1000 tweets in Kannada language
4	Sentiment Analysis of Restaurant online reviews	Li et al. (2023)	Conditional Survival Forest Model	Yelp
5	Sentiment Analysis of Student feedback	Dervenis et al. (2024)	ANN and RF, VADER	Student Feedback dataset
6	Trending Hashtags and associated Sentiment analysis	Bisht et al. (2024)	Naive Bayes, Gradient Boosting Decision Tree, and RF, VADER, TextBlob	Twitter Dataset
7	Hotel Rating Prediction from Sentiment Analysis of Reviews	Kumar et al. (2024)	LR, DT, SGD, RF, SVC, KNN	Hotelrec
8	Comment Classification	Rahman et al. (2024)	RF + Adaboost + SVM + soft-voting	IMDB, Twitter, Spambase
9	Sentiment Analysis of Android Applications Reviews	Qureshi et al. (2023)	LR, RF, MNB, KNN, and SVM	App reviews Dataset
10	Sentiment Analysis of Movie Reviews	Rahman et al. (2019)	Hybrid CNN-LSTM	IMDB movie review dataset and Amazon movie reviews dataset
11	Sentiment Analysis of Online Opinions	Jelodar et al. (2020)	Latent Dirichlet Allocation (LDA)- Bi-LSTM	Reddit Comments

(continued)

Table 1 (continued)

Sr. No	Context	Authors	Methodology used	Dataset
12	Sentiment Analysis of Online Reviews	Huang et al. (2022)	Lexicon-Based Sentiment CNN	Movie Reviews dataset, Sentiment labelled Sentences dataset
13	Personalised e-commerce Recommendation	Karabila et al. (2023)	Hybrid CF (Userbased, Item-based) BERT Fine-tuned	Amazon Musical Instruments reviews dataset
14	COVID-19 Sentiment Classification	Arbane et al. (2023)	Bi-LSTM	Twitter Dataset and Reddit Dataset
15	Amazon Product Reviews Classification	Anbumani et. al. (2023)	CNN	Amazon review dataset

and interpretability of results, ultimately enhancing the utility of sentiment analysis for business intelligence.

3 Proposed System

The proposed architecture, as illustrated in Fig. 1, includes feature engineering (text preprocessing, and word embedding), data balancing and feature extraction. The details of these modules are as follows:

3.1 Feature Engineering

The stage consists of four sub parts namely Text Preprocessing, and Word Embedding.

3.1.1 Text Preprocessing

In Text Preprocessing, classes are encoded in the number format. Positive will be encoded as 1 and Negative will be encoded as 2. Encoding the target variable is a necessary step in data preprocessing for ML and statistical modeling. It ensures that the target variable, whether categorical or textual, is represented in numeric form,

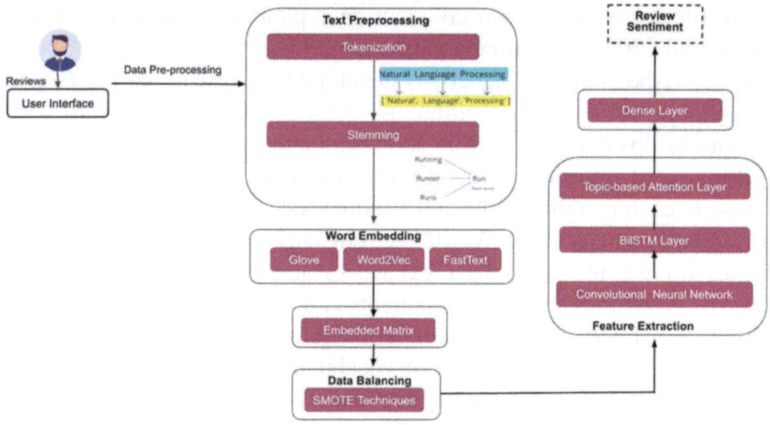

Fig. 1 The proposed architecture

allowing it to be compatible with ML algorithms, mathematical operations, model training, and evaluation metrics. This transformation maintains data consistency and enables the modeling process, making it an essential part of building effective predictive models. The choice of encoding method depends on the nature of the target variable, whether it has an ordinal or nominal relationship with the categories. Then, data is forwarded to Tokenizer.

Tokenization is important pre-processing step in NLP that consists of dividing a text into words or sub-words, referred to as tokens (Samuel & Krishna, 2024; Susmitha & Rajesh, 2023). These tokens serve as the primary building component for various NLP tasks, allowing computers to comprehend and process human language effectively. The obtained tokens are forwarded for the next step of pre-processing, namely, Stemming. Stemming that involves the reduction of words to their root or primitive form, known as "stem." The purpose of stemming is to simplify the text data by removing variations of words, thereby result in the dimensionality reduction of text while preserving its core meaning.

3.1.2 Word Embedding

It is a crucial technique in the field of NLP that bridges the gap between textual data and numerical representations, enabling machines to comprehend and process human language. It's a method in which the words are depicted with the help of dense words in a continuous vector space. The words that have similar meanings will be depicted with similar representations in vector form. In this research work, GloVe, Word2Vec and FastText word embeddings are used.

GloVe, which stands for Global Vectors for Word Representation aims to create dense vector representations for words on the basis of the semantic association between words according to their co-occurrence patterns in large text dataset (Abd El-Mageed et al., 2024; Suhartono et al., 2023).

Word2Vec (Abdelhady et al., 2024) developed by researchers at Google with the primary objective to create continuous vector spaces from the words where the spatial relationships between vectors show semantic relationships between words. The resulting word embeddings represent the meaning-based relationships between words close to each together in the embedding space.

FastText is an extension of Word2Vec, but unlike Word2Vec, which generates embeddings for whole words, FastText works at the sub-word level, meaning it breaks words down into smaller components called character n-grams (Abdelhady et al., 2024; Albahli & Nawaz, 2023). This technique enables FastText to create embeddings for the words that are out-of-vocabulary by aggregating the embeddings of their constituent character n-grams. Its ability to manage unfamiliar words makes it effective for languages especially the languages that have rich morphology or for tasks where the vocabulary is not fixed.

3.2 Feature Extraction Using Attention Based CNN-BiLSTM

To capture the complex patterns and context of textual reviews, a topic attention-based CNN-BiLSTM architecture is proposed for sentiment analysis. This hybrid approach combines the strengths of Convolution layers and BiLSTM network to capture local features as well as learn sequential and contextual dependencies bidirectionally. Additionally, the topic-based attention mechanism further enhances the model by focusing on most relevant parts of the text, ensuring high level feature extraction.

3.2.1 CNN

In recent years, CNN found numerous applications in modeling sequence problems like sentiment analysis, and text classification. It is a feedforward neural network that extract local patterns in the text using convolution and pooling layers. The convolution layer serves as the initial stage of the model, leveraging convolutional operations for feature extraction. Pooling layer reduces the number of dimensions spatially. This layer is adept at discerning local patterns within the input data, making it particularly valuable for tasks where recognizing specific features is essential. In our architecture, this layer is applied to the word embedding vector. Let the embedded matrix X $\in \mathbb{R}^{n \times d}$, where n represents number of words in the sentence and d represents the dimension of the word embeddings. A filter of size k slides over the word embedding matrix as $f_i = \sigma(W.X_{i:i+k-1} + b_c)$ with ReLU activation function. The term $X_{i:i+k-1}$ represents sub-matrix corresponds to sliding window, W as learnable weight matrix of the filter and b_c is bias term. After convolution operation, the feature maps are downsample to retain the most important features using mx-pooling layer. Finally, the output is then fed to the BiLSTM layer to capture sequence dependencies in both directions.

3.2.2 BiLSTM Layer

The BiLSTM layer process the input sequence in forward as well as backward direction (Fudholi et al., 2022; Meng et al., 2021), enables the model to capture long-term dependencies within the input text using bidirectional processing as depicted in Fig. 2. In the context of NLP, where understanding the contextual information of words is paramount, the BiLSTM layer proves instrumental by considering both past and future context at each time step. Basically, the LSTM addresses the issue of vanishing gradient error. It has three gates: input gate, forgot gate and output gate. Initially, the input from CNN layer x_t is passed through forget gate to determine the information to be discarded from the context. Mathematically, it can be expressed as

$$d_t = \sigma(W_d.x_t + U_d.h_{t-1} + b_d) \tag{1}$$

where d_t represents the degree of information retention in the input context. Then, the data flows into update gate to determine the information to update as

$$\mathrm{I}_t = \sigma(W_\mathrm{I}.x_t + U_\mathrm{I}.h_{t-1} + b_\mathrm{I}) \tag{2}$$

$$C_t = d_t \odot C_{t-1} + \mathrm{I}_t \odot \tanh(W_C.x_t + U_C.h_{t-1} + b_C) \tag{3}$$

Finally, the output is extracted from output gate as sequence features as

$$o_t = \sigma(W_o.x_t + U_o.h_{t-1} + b_o) \tag{4}$$

$$h_t = o_t \odot \tanh(C_t) \tag{5}$$

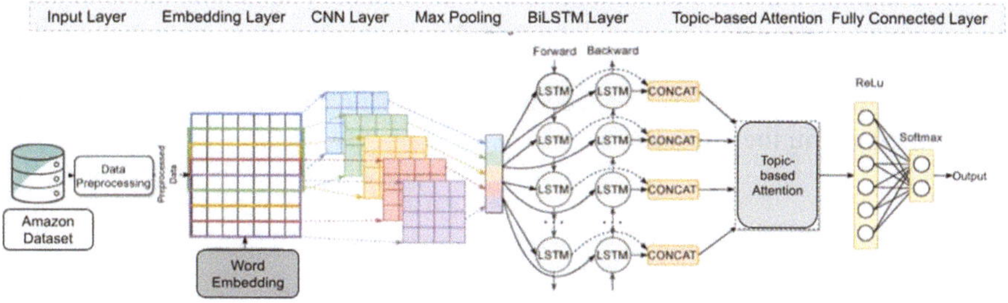

Fig. 2 A topic-based attention mechanism with a CNN-BiLSTM architecture

The output gate reflects the amount of information from internal memory at the output end (Tam et al., 2021).

Though LSTM resolves the issue of vanishing gradient error as well as the issue of long-term dependencies, the use of contextual information as post word information from the text still poses a challenge as the sentence is read only in one direction. To solve this problem, two-way information is extracted from the input sequence using BiLSTM which consist of forward as well as backward LSTM. Mathematically,

$$h_t^F = LSTM\left(x_t, h_{t-1}^F\right) \tag{6}$$

$$h_t^B = LSTM\left(x_t, h_{t-1}^B\right) \tag{7}$$

$$h_t = h_t^F \oplus h_t^B \tag{8}$$

where $\oplus$ represents the concatenation operator.

3.2.3 Topic-Based Attention Mechanism

This attention mechanism is introduced to assign weights to the words based on their relevance to the topics. Let $T \in \mathbb{R}^{N\times d}$ represents the topic matrix with N represents the number of topics and d is the dimension of word embeddings. The relevance score of each word corresponding to the topics is calculated using hidden state h_t from BiLSTM layer and the topic matrix T as

$$r_{t,n} = \cos(h_t, T_n) = \frac{h_t.T_n}{\|h_t\|\|T_n\|} \tag{9}$$

where T_n represents the nth topic vector and $r_{t,n}$ is the relevance score of word t to topic n. The resultant topic-relevance score matrix $R \in \mathbb{R}^{m\times N}$, where m denotes the number of words. To aggregate the topic-relevance scores into a single attention score, let $w_n \in \mathbb{R}^N$ weighing the topics contribution as

$$e_t = w^T R_t \tag{10}$$

where $R_t \in \mathbb{R}^N$ is the topic relevance vector for word t. Finally, the attention scores across all words in the sequence as

$$a_t = \frac{\exp(e_t)}{\sum_{t'} \exp(e_{t'})} \tag{11}$$

The context vector is calculated by aggregating the hidden states weighted by the attention score as

$$C = \sum_t a_t h_t \tag{12}$$

3.2.4 Fully Connected Layer:

The content vector is passed through fully connected layer as

$$Y = soft\ \max(WC + b) \tag{13}$$

where W and b represents the weight matrix and bias term for the output layer.

4 Experimentation

4.1 Dataset

This research work utilized Amazon Online Reviews dataset and Yelp dataset (Ayetiran, 2022). The dataset of Amazon online reviews for mobile phones, comprising over 400,000 reviews. This dataset consists of consumer opinions and sentiment regarding a diverse range of mobile phone products. Some of the features that are extracted for the purpose of this research described in Table 2.

Another dataset of Yelp reviews is taken that provides a rich collection of consumer opinions and sentiments regarding a diverse range of businesses. It includes 10,000 reviews present in the dataset. Some of the features that are extracted for the purpose of this research are in Table 3.

To train our models, we use Google Colab, with its GPU capabilities for accelerated computations. We also use Keras library with TensorFlow serving for python coding. During training, a mini-batch size of set as 32 and stochastic gradient descent over shuffled mini-batches is employed.

Table 2 Features of amazon dataset

Index	Feature	Description
1	Product name	The name of the mobile phone
2	Brand name	The brand of the mobile phone
3	Rating	The star rating provided by the reviewer
4	Review	The written review provided by the reviewer in text
5	Helpful votes	Number of users who found the review helpful

Table 3 Features of yelp dataset

Index	Feature	Description
1	Business ID	The unique id of the business in text
2	Stars	Number of stars (rating) given by user out of 5
3	Text	The textual review given by the user
4	Useful	Number of useful votes in the review

4.2 Data Balancing

The Amazon dataset that is being used in this research work has 284,997 positive reviews and 97,078 negative reviews. This results in an imbalanced dataset with models showing bias towards positive reviews. In the similar way, Yelp data set is also biased in one direction. To overcome this class imbalance problem, we incorporate Synthetic Minority Over-sampling Technique (SMOTE) (Al Habesyah et al., 2024; Habbat et al., 2023).

The basic idea of this technique is to create synthetic data for the minority class in order to reduce the imbalance. SMOTE works by finding the minority classes and then its neighbours are computed using K nearest neighbour technique. Then, new types are generated by interpolating between the selected class example and its neighbours. These steps are repeated until the dataset achieves the required balance. It is formulated as

$$S(x_i, x_{new}, \alpha) = x_i + \alpha(x_{new} - x_i) \tag{14}$$

where x_i is a minority class sample, x_{new} is a synthetic sample generated using SMOTE, α is a random value between 0 and 1, controlling the amount of over-sampling. Borderline-SMOTE is another technique that creates synthetic samples near decision boundaries that enhance classification accuracy but potentially causing class overlap (Abdulganiyu et al., 2025; Zhang & Liu, 2022). SVMSMOTE combines SVM and SMOTE that constructs synthetic samples for the minority class near the decision boundary (Börekci & Sevli, 2024; Nguyen et al., 2011).

4.3 Performance Metrics

We utilized several performance metrics to evaluate the performance of the proposed model.

1. ***Precision:*** It evaluates the accuracy of positive predictions and determined as

$$Precision = \frac{True\ Positives}{True\ Positives + False\ Positives} \tag{15}$$

A high value of precision reveals the lower value of false positives which makes it suitable for tasks where minimizing false positives is crucial.

2. ***Recall (Sensitivity):*** It evaluates the ability of the model to capture all positive class instances correctly and determined as

$$Recall = \frac{True\,Positives}{True\,Positives + False\,Negatives} \tag{16}$$

3. ***F1 Score:*** It is a metric which is quite helpful in datasets that have unbalanced class distribution and computed as

$$F1Score = 2 * \frac{Precision * Recall}{Precision + Recall} \tag{17}$$

4. ***Accuracy:*** It measures the ratio of predictions that are correct out of the total predictions and it is defined as

$$Precision = \frac{Correct\,Predictions}{Total\,Predictions} \tag{18}$$

5 Results and Discussion

Firstly, the performance of the proposed model for both the datasets is compared for three oversampling methods including SMOTE, Borderline SMOTE and SVMSMOTE using GloVe word embedding technique. Table 4 indicates that SMOTE method gives the best result with an accuracy of 96 and 97% for both Amazon dataset and Yelp dataset respectively followed by SVMSMOTE and Borderline SMOTE.

Furthermore, the comparative analysis of performance metrics, as illustrated in Tables 5 and 6, provides a comprehensive understanding of the efficacy of different word embeddings: GloVe, Word2Vec, and FastText—across two distinct datasets: Amazon and Yelp taking SMOTE data balancing techniques.

Table 4 Performance Evaluation with different oversampling methods

Dataset	Oversampling technique	Accuracy
Amazon dataset	SMOTE	0.96
	Borderline SMOTE	0.93
	SVMSMOTE	0.94
Yelp dataset	SMOTE	0.97
	Borderline SMOTE	0.94
	SVMSMOTE	0.95

Table 5 Performance evaluation with amazon dataset

Embedding	Model	Performance metrics			
		Accuracy	Recall	F1 Score	Precision
GloVe	CNN	0.92	0.90	0.91	0.93
	LSTM	0.89	0.88	0.89	0.90
	BiLSTM	0.91	0.89	0.90	0.92
	LSTM + BiLSTM	0.92	0.90	0.91	0.93
	CNN + LSTM	0.95	0.92	0.92	0.94
	CNN + BiLSTM	0.95	0.92	0.92	0.94
	Proposed Model	**0.96**	**0.93**	**0.93**	**0.94**
Word2Vec	CNN	0.88	0.85	0.86	0.89
	LSTM	0.87	0.86	0.87	0.88
	BiLSTM	0.89	0.87	0.88	0.90
	LSTM + BiLSTM	0.92	0.90	0.91	0.93
	CNN + LSTM	0.95	0.92	0.92	0.94
	CNN + BiLSTM	0.95	0.92	0.92	0.94
	Proposed Model	**0.93**	**0.90**	**0.90**	**0.91**
FastText	CNN	0.86	0.83	0.84	0.88
	LSTM	0.85	0.84	0.85	0.86
	BiLSTM	0.87	0.85	0.86	0.88
	LSTM + BiLSTM	0.92	0.90	0.91	0.93
	CNN + LSTM	0.95	0.92	0.92	0.94
	CNN + BiLSTM	0.95	0.92	0.92	0.94
	Proposed Model	**0.92**	**0.89**	**0.89**	**0.90**

GloVe consistently emerges as the top performer across both datasets, showcasing remarkable accuracy, recall, F1 Score, and precision values. Particularly noteworthy is the Proposed Model utilizing GloVe embeddings, achieving an impressive accuracy of 96% on the Amazon dataset and 97% on the Yelp dataset. These results underscore the robust capability of GloVe in accurately capturing semantic relationships, making it a compelling choice for various natural language processing tasks.

While Word2Vec doesn't reach the pinnacle set by GloVe, it consistently establishes itself as the second-best performer, with a notable accuracy of 93% in the proposed Model on the Amazon dataset and 95% on the Yelp dataset, highlights Word2Vec's versatility. Its competitive recall, F1 Score, and precision values reinforce its suitability for a wide range of NLP applications.

FastText, although trailing behind GloVe and Word2Vec, presents itself as a viable option. The updated results for the proposed model on the Yelp dataset reveal an improved accuracy of 92%, emphasizing its adaptability in scenarios where computational efficiency is a primary concern.

Table 6 Performance evaluation with yelp dataset

Embedding	Model	Performance metrics			
		Accuracy	Recall	F1 Score	Precision
GloVe	CNN	0.92	0.90	0.91	0.93
	LSTM	0.89	0.88	0.89	0.90
	BiLSTM	0.91	0.89	0.90	0.92
	LSTM + BiLSTM	0.92	0.90	0.91	0.93
	CNN + LSTM	0.95	0.92	0.92	0.94
	CNN + BiLSTM	0.95	0.92	0.92	0.94
	Proposed Model	**0.97**	**0.94**	**0.94**	**0.95**
Word2Vec	CNN	0.86	0.83	0.84	0.88
	LSTM	0.85	0.84	0.85	0.86
	BiLSTM	0.87	0.85	0.86	0.88
	LSTM + BiLSTM	0.90	0.88	0.89	0.91
	CNN + LSTM	0.92	0.89	0.90	0.92
	CNN + BiLSTM	0.92	0.89	0.90	0.92
	Proposed Model	**0.95**	**0.92**	**0.92**	**0.94**
FastText	CNN	0.78	0.75	0.76	0.80
	LSTM	0.80	0.78	0.79	0.81
	BiLSTM	0.82	0.80	0.81	0.83
	LSTM + BiLSTM	0.85	0.82	0.83	0.86
	CNN + LSTM	0.88	0.85	0.86	0.89
	CNN + BiLSTM	0.89	0.86	0.87	0.90
	Proposed Model	**0.92**	**0.89**	**0.90**	**0.91**

Moreover, Fig. 3 illustrates ROC curves for both the datasets with GloVe word embedding technique and SMOTE method. The proposed model gives ROC value 0.94 for Amazon dataset and 0.95 for Yelp dataset.

(a) Amazon Dataset (b) Yelp Dataset.

In conclusion, the comparative analysis underscores the importance of tailoring word embedding choices to the specific requirements of a given natural language processing task. GloVe stands out as a preferred choice for applications demanding high precision and recall, while Word2Vec provides an optimal balance between accuracy and computational efficiency. FastText, though not achieving the same performance levels, remains a viable option in situations where rapid computation takes precedence. These insights provide valuable guidance for practitioners aiming to optimize their choice of word embeddings based on the nuances of their specific applications.

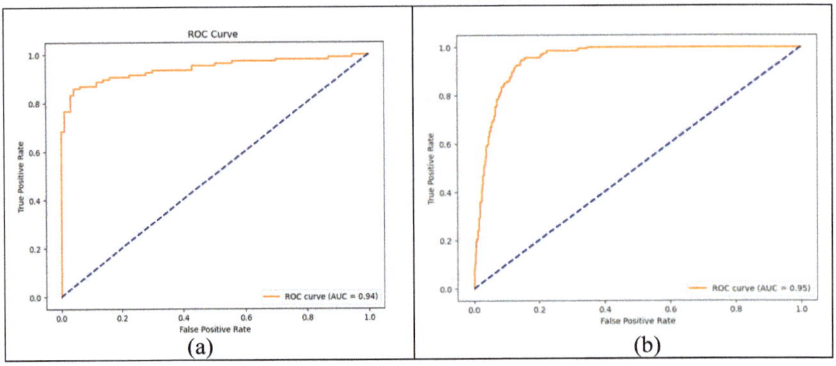

Fig. 3 ROC curves with GloVe word embedding technique and SMOTE for

6 Deployment and Real-Time Inference

After training the ML model, we save it in the.h5 format, a widely used format for storing neural network models, using a framework such as TensorFlow or Keras. This ensures that the model's architecture, weights, and configuration are preserved for future use. Following model training and saving, we employ the FastAPI framework to facilitate real-time inference. FastAPI provides a simple and efficient way to create API endpoints, enabling seamless communication between the trained model and external applications.

6.1 FastAPI

FastAPI is employed to create an API endpoint dedicated to making predictions. The endpoint is configured to receive input data, process it through the loaded model, and return the predictions. This FastAPI endpoint provides a means for external applications, such as a frontend user interface, to send input data and receive real-time predictions from the model.

6.2 Sending Results to Frontend

To complete the real-time inference process, a frontend application can make HTTP-POST requests to the FastAPI endpoint. Figure 4 shows the user interface. This interaction allows the frontend to dynamically obtain predictions based on user input.

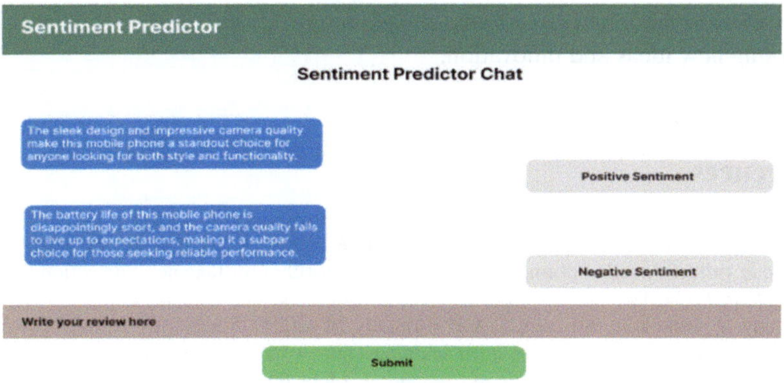

Fig. 4 User interface

7 Conclusion and Future Directions

This research work put forth a comprehensive examination into the domain of sentiment analysis of online reviews using a wide range of embedding methods and learning networks. The proposed framework employs an amalgamation of CNN, BiLSTM and topic level attention and compares three keyword embedding techniques namely GloVe, Word2Vec, and FastText. Moreover, the performance of three class oversampling techniques- SMOTE, Borderline SMOTE and SVMSMOTE is compared. The proposed model stands out as a strong performer across both the datasets namely Amazon and Yelp with high values of performance metrics including accuracy, recall, F1 Score, and precision.

The comparative analysis shows SMOTE gives better accuracy in the proposed model followed by Borderline SMOTE and SVMSMOTE with both the datasets. The analysis also points to the importance of suitable word embedding technique for sentiment analysis. GloVe performs better in terms of precision and recall, while Word2Vec gives an optimal balance between accuracy and computational efficiency. FastText proves to be a viable option where rapid computation is required.

To facilitate the smooth communication of the proposed model with external applications, FastAPI framework is used after saving it in the.h5 format. Prospective directions for future work include:

- Exploring more Embeddings in the field trained to get more accurate model for sentiment analysis in specialized fields.
- Making models more interpretable to assist users in recognizing the factors that affect sentiment classifications.
- Examining the adaptability of model by checking its performance in dynamic online review environments.
- Extending the study to include multiple modes of data which includes both text and visual information in order to further refine sentiment analysis models, and extending its versatility.

The above-mentioned directions will help future advancement in the field of NLP, promoting new ideas and innovation.

References

Abd El-Mageed, A. A., Abohany, A. A., Ali, A. H., & Hosny, K. M. (2024). An adaptive hybrid African vultures-aquila optimizer with Xgb-Tree algorithm for fake news detection. *Journal of Big Data, 11*(1), 41.

Abdelhady, N., Hassan A. Soliman, T., & F. Farghally, M. (2024). Stacked-CNN-BiLSTM-COVID: An effective stacked ensemble deep learning framework for sentiment analysis of Arabic COVID-19 tweets. *Journal of Cloud Computing, 13*(1), 85.

Abdulganiyu, O. H., Tchakoucht, T. A., Saheed, Y. K., & Ahmed, H. A. (2025). XIDINTFL-VAE: XGBoost-based intrusion detection of imbalance network traffic via class-wise focal loss variational autoencoder. *The Journal of Supercomputing, 81*(1), 1–38.

Al Habesyah, N. Z., Herteno, R., Indriani, F., Budiman, I., & Kartini, D. (2024). Sentiment analysis of TikTok shop closure in Indonesia on Twitter using supervised machine learning. *Journal of Electronics, Electromedical Engineering, and Medical Informatics, 6*(2), 148–156.

Albahli, S., & Nawaz, M. (2023). TSM-CV: Twitter sentiment analysis for COVID-19 vaccines using deep learning. *Electronics, 12*(15), 3372.

Anbumani, P., & Selvaraj, K. (2023) Enhancing sentiment analysis classification for amazon product reviews using CNN-sigTan-Beta activation function. *Multimedia Tools and Applications, 19*, 1–18.

Arbane, M., Benlamri, R., Brik, Y., & Alahmar, A. D. (2023). Social media-based COVID-19 sentiment classification model using Bi-LSTM. *Expert Systems with Applications, 212*, Article 118710.

Ayetiran, E. F. (2022). Attention-based aspect sentiment classification using enhanced learning through CNN-BiLSTM networks. *Knowledge-Based Systems, 252*, Article 109409.

Barbado, R., Araque, O., & Iglesias, C. A. (2019). A framework for fake review detection in online consumer electronics retailers. *Information Processing & Management, 56*(4), 1234–1244.

Baziotis, C., Pelekis, N., & Doulkeridis, C. (2017). Data stories at SemEval 2017 task 4: Deep LSTM with attention for message-level and topic based sentiment analysis. In *Proc. 11th Int. Workshop Semantic Eval. (SemEval)* (pp. 747–754).

Bisht, T., Singh, M., & Singhal, S. (2024). Determining Twitter trending hashtags and sentiments associated via machine learning approaches. *SN Computer Science, 5*(8), 1002.

Börekci, A., & Sevli, O. (2024). A classification study for Turkish folk music makam recognition using machine learning with data augmentation techniques. *Neural Computing and Applications, 36*(4), 1621–1639.

Chehal, D., Gupta, P., & Gulati, P. (2021). Implementation and comparison of topic modeling techniques based on user reviews in e-commerce recommendations. *Journal of Ambient Intelligence and Humanized Computing, 12*, 5055–5070.

Dervenis, C., Kanakis, G., & Fitsilis, P. (2024). Sentiment analysis of student feedback: A comparative study employing lexicon and machine learning techniques. *Studies in Educational Evaluation, 83*, Article 101406.

Du, J., Gui, L., He, Y., Xu, R., & Wang, X. (2019). Convolution-based neural atten- tion with applications to sentiment classification. *IEEE Access, 7*, 22983–27992.

Fudholi, D. H., Nayoan, R. A. N., & Rani, S. (2022). Stock prediction based on Twitter sentiment extraction using BiLSTM-attention. *Indonesian Journal of Electrical Engineering and Informatics (IJEEI), 10*(1), 187–198.

Garcıa-Moya, L., Anaya-S´anchez, H., & Berlanga-Llavori, R. (2013, May/June). Retrieving product features and opinions from customer reviews. *IEEE Computer Society*, 19–27.

Habbat, N., Nouri, H., Anoun, H., & Hassouni, L. (2023). Sentiment analysis of imbalanced datasets using BERT and ensemble stacking for deep learning. *Engineering Applications of Artificial Intelligence, 126*, Article 106999.

Huang, F., Li, X., Yuan, C., Zhang, S., Zhang, J., & Qiao, S. (2021). Attention-emotion-enhanced convolutional LSTM for sentiment analysis. *IEEE Transactions on Neural Networks and Learning Systems, 33*(9), 4332–4345.

Huang, M., Xie, H., Rao, Y., Liu, Y., Poon, L. K. M., & Wang, F. L. (2022). Lexicon-based sentiment convolutional neural networks for online review analysis. *IEEE Transactions on Affective Computing, 13*(3), 1342–1348.

Jain, P. K., Srivastava, G., Lin, J. C. W., & Pamula, R. (2022). Unscrambling customer recommendations: A novel lstm ensemble approach in airline recommendation prediction using online reviews. *IEEE Transactions on Computational Social Systems, 9*(6), 1777–1784.

Jelodar, H., Wang, Y., Orji, R., & Huang, S. (2020). Deep sentiment classification and topic discovery on novel coronavirus or COVID-19 online discussions: NLP using LSTM recurrent neural network approach. *IEEE Journal of Biomedical and Health Informatics, 24*(10), 2733–2742.

Jiang, W., Zhou, K., Xiong, C., Du, G., Ou, C., & Zhang, J. (2023). KSCB: A novel unsupervised method for text sentiment analysis. *Applied Intelligence, 53*(1), 301–311.

Karabila, I., Darraz, N., EL-Ansari, A., Alami, N., & EL Mallahi, M. (2023). BERT-enhanced sentiment analysis for personalized e-commerce recommendations. *Multimedia Tools and Applications*, 1–26.

Khedkar, S., & Shinde, S. (2020) Linguistic feature-based praise or complaint classification from customer reviews. In *Intelligent computing, information and control systems: ICICCS 2019* (pp. 470–481). Springer International Publishing.

Kim, R. Y. (2021). Using online reviews for customer sentiment analysis. *IEEE Engineering Management Review, 49*(4), 162–168.

Kumar, M., Kumar, C., Kumar, N., & Kavitha, S. (2024). Efficient hotel rating prediction from reviews using ensemble learning technique. *Wireless Personal Communications, 137*(2), 1161–1187.

Lei, X., Qian, X., & Zhao, G. (2016). Rating prediction based on social sentiment from textual reviews. *IEEE Transactions on Multimedia, 18*(9), 1910–1919.

Li, H., Bruce, X. B., Li, G., & Gao, H. (2023). Restaurant survival prediction using customer-generated content: An aspect-based sentiment analysis of online reviews. *Tourism Management, 96*, Article 104707.

Lin, Y., Li, J., Yang, L., Xu, K., & Lin, H. (2019). Sentiment analysis with comparison enhanced deep neural network. *IEEE Access, 8*, 68363–68371.

Meng, J., Dong, Y., Long, Y., & Zhao, D. (2021). An attention network based on feature sequences for cross-domain sentiment classification. *Intelligent Data Analysis, 25*(3), 627–640.

Murugaiyan, S., & Uyyala, S. R. (2023). Aspect-based sentiment analysis of customer speech data using deep convolutional neural network and bilstm. *Cognitive Computation, 15*(3), 914–931.

Nguyen, H. M., Cooper, E. W., & Kamei, K. (2011). Borderline over-sampling for imbalanced data classification. *International Journal of Knowledge Engineering and Soft Data Paradigms, 3*(1), 4–21.

Obiedat, R., Qaddoura, R., Ala'M, A.Z., Al-Qaisi, L., Harfoushi, O., Alrefai, M. A., & Faris, H. (2022). Sentiment analysis of customers' reviews using a hybrid evolutionary SVM-based approach in an imbalanced data distribution. *IEEE Access, 10*, 22260–22273.

Patil, P. B., Ijeri, D., Kulkarni, S. A., Burkaposh, S. S., Bhuyyar, R., & Gugawad, V. (2023). Comparative study of machine learning algorithms for Kannada twitter sentimental analysis. *Multimedia Tools and Applications, 83*, 1–21.

Qureshi, A. A., Ahmad, M., Ullah, S., Yasir, M. N., Rustam, F., & Ashraf, I. (2023). Performance evaluation of machine learning models on large dataset of android applications reviews. *Multimedia Tools and Applications, 82*(24), 37197–37219.

Rae Yule Kim. (2021). Using online reviews for customer sentiment analysis. *IEEE Engineering Management Review, 49*(4), 162–168.

Rahman, M. M., Shiplu, A. I., & Watanobe, Y. (2024). Comment Class: A robust ensemble machine learning model for comment classification. *International Journal of Computational Intelligence Systems, 17*(1), 184.

Rehman, A. U., Malik, A. K., Raza, B., & Ali, W. (2019). A hybrid CNN-LSTM model for improving accuracy of movie reviews sentiment analysis. *Multimedia Tools and Applications, 78*, 26597–26613.

Samuel, P., & Krishna, B. V. (2024). Exploration of sentiment analysis in twitter propaganda: A deep dive. *Multimedia Tools and Applications, 83*(15), 44729–44751.

Shen, Y., Tan, S., Sordoni, A., & Courville, A. (2018). O*rdered neurons: Integrating tree structures into recurrent neural networks*. arXiv:1810.09536. [Online]. Available: http://arxiv.org/abs/1810.09536

Shin, B., Lee, T., & Choi, J. D. (2016). *Lexicon integrated CNN models with attention for sentiment analysis*. arXiv preprint arXiv:1610.06272.

Suhartono, D., Purwandari, K., Jeremy, N. H., Philip, S., Arisaputra, P., & Parmonangan, I. H. (2023). Deep neural networks and weighted word embeddings for sentiment analysis of drug product reviews. *Procedia Computer Science, 216*, 664–671.

Susmitha, M., & Rajesh, P. (2023) Information extraction with two-layered ODNN and semantic analysis for opinion mining. *Multimedia Tools and Applications, 83*(16), 1–29.

Tam, S., Said, R. B., & Tanriöver, Ö. Ö. (2021). A ConvBiLSTM deep learning model-based approach for Twitter sentiment classification. *IEEE Access, 9*, 41283–41293.

Ullal, M. S., Spulbar, C., Hawaldar, I. T., Popescu, V., & Birau, R. (2021) The impact of online reviews on e-commerce sales in India: A case study. *Economic Research-EkonomskaIstraživanja, 34*(1), 2408–2422.

Wang, Y., Huang, M., Zhu, X., & Zhao, L. (2016). Attention-based LSTM for aspect-level sentiment classification. In *Proc. Conf. Empirical Methods Natural Lang. Process* (pp. 606–615).

Xia, H., Yang, Y., Pan, X., Zhang, Z., & An, W. (2020). Sentiment analysis for online reviews using conditional random fields and support vector machines. *Electronic Commerce Research, 20*, 343–360.

Yang, Z., Li, Q., Charles, V., Xu, B., & Gupta, S. (2023). *Online product decision support using sentiment analysis and fuzzy cloud-based multi-criteria model through multiple e-commerce platforms*. IEEE Transactions on Fuzzy Systems.

Yin, W., Schütze, H., Xiang, B., & Zhou, B. (2016). Abcnn: Attention-based convolutional neural network for modeling sentence pairs. *Transactions of the Association for Computational Linguistics, 4*, 259–272.

Zhang, Y., & Liu, Q. (2022). On IoT intrusion detection based on data augmentation for enhancing learning on unbalanced samples. *Future Generation Computer Systems, 133*, 213–227.

The Financial Services Lens

Quantum Machine Learning for Stock Price Movement Prediction

Manoranjan Gandhudi, P. J. A. Alphonse, and G. R. Gangadharan

Abstract Stock price prediction has been a challenging task in financial markets. In recent years, the integration of natural language processing methods with machine learning algorithms has gained significant attention for its potential in extracting valuable insights from text data such as news articles. Classical machine learning algorithms often face exponential complexity with growing datasets and struggle with escaping local optima, while quantum machine learning techniques leverage quantum parallelism and interference effects to potentially offer exponential speedup and more efficient exploration of high-dimensional spaces. This chapter uses an approach combining the linguistic patterns from news articles headlines with cutting edge quantum machine learning techniques for stock price prediction. Textual features are extracted from news articles through Bag of Words, TF-IDF, and N-gram representations, aiming to encapsulate the semantic meaning and contextual variations of the articles. The integration of quantum techniques with textual features achieves superior prediction accuracy, representing the quantum machine learning importance in financial forecasting tasks. The experimental outcomes demonstrate the quantum random forest effectiveness in predicting stock prices compared to other quantum machine learning methods. Based on the outcomes, we observe that the quantum random forest method provides an accuracy of 96% and a precision of 97%. Furthermore, the study highlights the significance of leveraging textual data and advanced computational techniques for enhancing stock market prediction models, paving the way for more robust and accurate financial decision-making systems.

Keywords Stock Price Prediction · Natural Language Processing · Quantum Machine Learning

---

M. Gandhudi · P. J. A. Alphonse · G. R. Gangadharan (✉)
National Institute of Technology, Tiruchirappalli, India
e-mail: ganga@nitt.edu

B. Basu et al. (eds.), *Data Analytics for Decision Making towards Business Excellence*, Uncertainty and Operations Research, https://doi.org/10.1007/978-981-95-2429-7_7

1 Introduction

Stock price prediction serves as a pivotal tool in the field of finance, enabling investors to navigate the stock market price complexities with greater insight and foresight (Das et al., 2024). By leveraging predictive models and algorithms, investors can arrive with informed decisions about holding, or selling the stocks, optimizing their portfolios for maximum returns while minimizing risks (Albahli et al., 2023). Moreover, accurate predictions aid in strategic planning and risk management, empowering individuals and institutions to align their financial goals with market dynamics (Liu et al., 2020). Beyond individual investment decisions, stock price predictions offer valuable insights into broader economic trends and market sentiment, shaping corporate strategies and influencing policy decisions (Xie et al., 2024). In essence, the capability of stock price prediction involves not only in its drive to financial success but also in its role as an economic health barometer and a catalyst for informed decision-making in the global financial landscape (Hu et al., 2024).

The impact of news article headlines on stock prices can be substantial, often leading to short-term fluctuations in market sentiment and trading activity. Headlines wield significant power in shaping investors' perceptions of a company, its products, or its prospects, thus influencing their buying or selling decisions (Valle-Cruz et al., 2022). Positive headlines, including robust earnings announcements or strategic partnerships, tend to propel stock prices higher as investors view the company favorably and anticipate future growth (Nekrasov et al., 2022). Conversely, negative headlines, like reports of regulatory scrutiny or declining sales, can prompt stock prices to decline as investors react to perceived risks or uncertainties. However, it is crucial to acknowledge that the news headlines influence stock market prices, based on factors such as the credibility of the source, the perceived importance of the news, and the prevailing market sentiment at the time of publication (Ashtiani & Raahemi, 2023). While news headlines can indeed impact long-term stock performance, short-term price movements are primarily driven by fundamental factors such as growth prospects, earnings, and overall market conditions (Ray et al., 2021). Therefore, while considering news article headlines, it is essential for understanding stock price volatility and investors should exercise caution and evaluate factors when making investment decisions in a broader range.

To perform stock price prediction, a diverse array of machine learning approaches is employed, each offering unique capabilities to model and forecast market dynamics (Kumbure et al., 2022). Linear regression provides a foundational approach by establishing linear relationships between historical data and future stock prices, albeit with limitations in capturing complex patterns. Support Vector Machines (SVM) excel in handling both linear and non-linear relationships through kernel functions, offering robust predictions even amidst intricate market data. Random Forest, with its ensemble learning mechanism, adeptly tackles noise and missing data while maintaining prediction accuracy. Gradient Boosting Machines (GBM) sequentially refine predictions, ensuring high precision while mitigating overfitting risks. Deep Learning

(DL) techniques including Convolutional Neural Networks (CNN) and Long Short-Term Memory (LSTM) excel in identifying temporal and spatial dependencies within stock data, providing unparalleled insights into market trends. Gated Recurrent Units (GRU) offer an efficient alternative to LSTM, proficiently modeling long-term dependencies. These algorithms, among others, form a comprehensive toolkit for analysts and researchers, each contributing uniquely to the quest for accurate stock price predictions in financial markets (Bansal et al., 2022).

There are several machine learning techniques often employed in predicting stock price movement. Each algorithm comes with its own set of strengths and weaknesses, and while claims are made about their performance, empirical evidence to support these claims can sometimes be lacking. For instance, one commonly used technique is linear regression. It is valued for its interpretability and simplicity, making it a popular option for establishing baseline models. However, its reliance on linear relationships can limit its accuracy, particularly in identifying the complex and distinct nonlinear patterns often found in stock data. Decision trees, another popular choice, offer interpretability and can handle nonlinear relationships better than linear regression. Yet, they are prone to overfitting, especially when dealing with high-dimensional or noisy data, which can lead to poor generalization performance. Ensemble methods like Random Forest and Gradient Boosting are favored for their capability to mitigate overfitting and improve prediction accuracy by combining multiple weak learners. However, they can be computationally expensive and need careful parameter tuning to arrive at optimal performance.

Moving on to neural networks, they excel at capturing complex patterns in data and have shown promising results in stock price prediction. However, they are often criticized for their black-box nature (Gandhudi et al., 2024b), making it challenging to interpret their decisions. Additionally, they may need a huge quantity of data and computational facilities to train in an effective way. Support Vector Machines (SVMs) are valued for their potential to handle high-dimensional data and distinct nonlinear relationships with kernel functions usage. However, they can be sensitive to the kernel and regularization parameters, which may impact their performance outcomes. Finally, time series models like Autoregressive Integrated Moving Average (ARIMA), Fuzzy Time Series (Didugu et al., 2024) and Long Short-Term Memory (LSTM) networks are especially designed for sequential data like stock prices. ARIMA models capture temporal dependencies and trends in the data but may struggle with nonlinear patterns. LSTM networks, on the other hand, excel at capturing long-term dependencies but can be challenging to train and prone to overfitting, especially with limited data.

Additionally, it could benefit from highlighting the inherent challenges associated with applying these algorithms to stock market data, including the presence of non-stationarity, noisy data, and the unpredictability of market behavior (Moews & Ibikunle, 2020). Furthermore, these algorithms do not address the potential drawbacks of relying solely on ML methods for stock price movement prediction, including the inability to account for exogenous events or market sentiment. Conventional machine learning algorithms often encounter challenges such as exponential complexity as datasets increase in size, and they may struggle to overcome local optima.

To overcome these limitations, in our work, we trained the data using quantum machine learning approaches to perform the stock price movement prediction. Regarding the quantum models significance in stock price prediction, quantum computing has the aptness to revolutionize financial modeling by performing complex computations at speeds incredible with classical computers. Quantum techniques can handle large amounts of data and explore potential scenarios of multitudes simultaneously, offering the potential for more accurate predictions and risk assessments. Quantum models have the potential to alleviate classical machine learning constraints, particularly in detecting nonlinear distinct relationships and handling high-dimensional data.

The chapter is organized as follows. Section 2 discusses a literature survey on stock price prediction using various sentiment analysis. Section 3 illustrates the proposed methodology, Sect. 4 presents the outcomes of the proposed method including comparative analysis with the other quantum models, Conclusion and future work included in Sect. 5.

2 Literature Survey

Cho et al. (2024) applied regression analysis to examine the effects of 503,107 news related to 1012 organizations on U.S. biotechnology and pharmaceutical organizations using CRSP daily stock return data and RavenPack news article data. They identified that in comparison to smaller organizations with a higher market capitalization typically have a lesser degree of abnormal returns. Maqbool et al. (2023) conducted an experiment utilizing a decade of historical stock data and financial related news from four individual organizations across various sectors to forecast the future stock trends. The maximum accuracy of 90 obtained for both trend and future trends over a time period of 10 days. Khan et al. (2022) performed a stock prediction utilizing various machine learning related classifiers using social media and news articles. Among the various machine learning classifiers, random forest is found consistent and provides accurate results. Hao et al. (2021) presented twin SVM with a fuzzy hyperplane to forecast stock price by merging the capabilities of SVM and fuzzy set theory. This method aids investors in uncertainty, risk, and confidence levels by offering accurate predictions on stock price direction, ultimately reducing unexpected risks.

Swathi et al. (2022) developed a hybrid model linking LSTM and learning and teaching based optimization techniques to evaluate the correlation between historical stock price and tweet data. The correlation between tweets is utilized to determine how user mood affects the stock price. Zhao et al. (Zhao & Yang, 2023) presented a hybrid framework combining an emotion convolutional neural network (ECNN), denoising autoencoder (DAE) methods, and LSTM to estimate stock price and simulation trading. Distinct emotion strengths are extracted using the exponential time function with a maximized performance in both risk and return. Mu et al. (2023) introduced a model named MS-SSA-LSTM (multi source Sparrow Search Algorithm based LSTM) to train and test six datasets from the Chinese financial market. They employed four assessment indicators to examine the predictive performance of the model. Li et al. (2020) proposed a tensor based multi modal driven LSTM approach to perform stock prediction based on online news. According to Li et al. (2020), the online news has a high impact on stock movements.

Yang et al. (Li & Pan, 2022) introduced a deep learning architecture (DLA) to anticipate stock prices by linking multiple recurrent neural networks in an ensemble learning model. Experiments were performed on distinct recurrent neural network techniques to evaluate the significance of the method. The model achieved a maximum of 67% accuracy on the hold-out test dataset in MDA. Li et al. (Li & Pan, 2022) proposed a novel DL model using a connecting ensemble learning technique to sequentially integrate two RNN's to forecast succeeding stock prices. Choi et al. (2023) discussed hybrid information mixing modules for stock price movement prediction, involving the mixing of time-series and semantic features. The interaction between data characteristics was strengthened through row-wise and column-wise operations. Wang et al. (2023) created a hybrid estimation model integrating sentiment analysis and technical indicators to forecast stock market trends for AMZN, GOOG, EBAY, AAPL, and C13. The model comprises three phases and achieved an improved accuracy and the F1-score for DJIA. This underscores the effectiveness of combining technical indicators and sentiment analysis in stock estimation. Wan et al. (2021) uses various NLP techniques to analyze the sentiment of various news on 87 industries for a period of 7 years. The work investigates the propagation of news sentiment in organization networks and evaluates the associated stock price movements including market volatility and price. Koratamaddi et al. (2021) proposed a deep reinforcement algorithm to manage the stock portfolio depending on the market sentiment analysis.

Ma et al. (2024) proposed VGC-GAN, a multi-graph convolutional adversarial framework for stock price prediction. It generates correlation graphs from historical data and employs a GAN with MSE loss, combining Multi-GCN and GRU for enhanced prediction. However, its limitation includes high computational requirements and sensitivity to graph quality, impacting generalizability. Wen et al. (2024) proposed a stock prediction model using multilevel wavelet interaction network, integrating Maximum Overlapping Discrete Wavelet Transform into deep learning, a module using deep difference to reduce autocorrelated errors, and an enhanced Hull Moving Average to capture market trends. However, the complexity of the model may hinder real-time application efficiency. Lu et al. (Lu & Xu, 2024) proposed a time-series RNN for stock prediction, integrating sliding windows and trading volume data to capture trends and turning points. The model enhances efficiency by compressing data and upgrading the price–volume relationship. However, its limitation lies in handling highly volatile market conditions effectively.

Gandhudi et al. (2024a) proposed a hybrid framework to predict stock market prices combining a genetic algorithm and quantum neural network and transformer models to compute sentiment scores from tweets, integrating them with stock data.

While the method improves prediction accuracy, its limitations include high computational complexity and potential overfitting due to model intricacy. Xu et al. (2024) proposed the Quantum Gramian Angular Field, combining quantum computing with deep learning for improved stock return forecasting. Empirical tests on datasets from China, Hong Kong, and the USA showed higher accuracy than traditional methods. However, the approach faces scalability challenges due to current quantum hardware limitations. Li et al. (2023) proposed a hardware-efficient Quantum RNN, constructed with staggered Quantum Recurrent Blocks (QRBs) to reduce the coherence time requirements for NISQ devices. Tested on meteorological, stock, and text data, it outperformed classical RNNs and QNNs. However, scalability to larger datasets remains a limitation. Kobayashi et al. (2023) investigated quantum and quantum ML algorithms for stock return prediction, comparing Quantum Neural Networks (QNNs) and tensor networks with classical models like linear regression and neural networks. Empirical results from the Japanese stock market showed tensor networks outperform classical benchmarks. However, QNNs yielded lower risk-adjusted returns over extended periods, highlighting challenges in stability.

Based on the literature review, we observe that conventional ML and DL techniques have advancements in stock price prediction based on news articles. However, these techniques have limitations in computational power, scalability, and optimization efficiency. These techniques struggle with handling huge datasets, complex models, and ensuring data privacy and security. Quantum algorithms, on the other hand, offer a promising alternative with the potential for exponential speedups and improved computational abilities. By harnessing quantum phenomena including superposition and entanglement, quantum techniques have the ability to address these limitations and handle complex problems in an effective way.

3 Proposed Methodology

The methodology implemented in this study is structured around a series of necessary steps to analyze and classify data efficiently. Beginning with data preprocessing to ensure quality and consistency, the process moves on to feature extraction, utilizing traditional techniques including bag of words, TF-IDF, and a bi-gram model to capture essential linguistic patterns. Quantum machine learning is then applied to effectively learn the underlying patterns of the data, leveraging quantum computing principles for optimization. Notably, the choice of a binary classification task underscores the practical relevance of the study, with a keen emphasis on computational efficiency and speed, ensuring applicability to real-world scenarios where rapid analysis is paramount. This methodology bridges classical and quantum computing approaches, offering a promising direction for data analysis and classification problems. The framework of this work is presented as follows (see Fig. 1).

Quantum Machine Learning (QML) offers significant advantages over classical Machine Learning (ML) by incorporating the principles of quantum mechanics

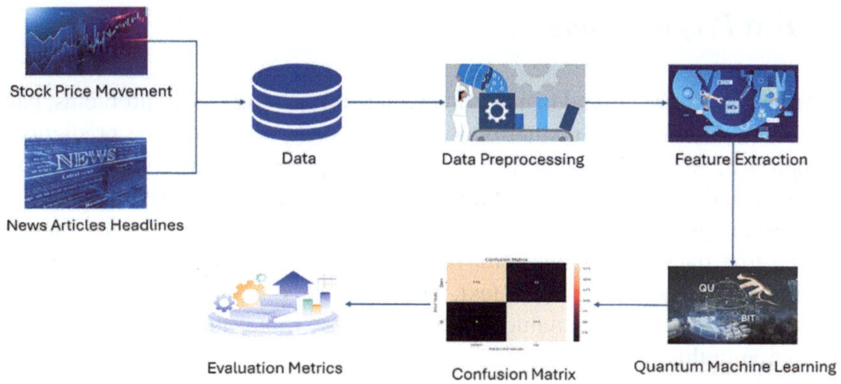

Fig. 1 Framework of the work

including entanglement, superposition, and quantum parallelism. These enable QML algorithms to explore multiple states simultaneously, allowing for exponentially faster computation in specific tasks compared to classical counterparts. While traditional ML relies on deterministic algorithms and processes data in classical bits (0s and 1s), QML uses quantum bits, which are involved in a superposition of states. This fundamental difference allows QML to handle high-dimensional data spaces more efficiently, making it specifically well suitable for complex optimization tasks, feature selection, and large-scale data modeling. Further, QML can exploit quantum entanglement to capture intricate relationships between variables that classical algorithms may struggle to model. However, the superiority of QML depends on the problem and current quantum hardware constraints, as its full potential remains contingent on advancements in quantum computing technology.

3.1 Dataset Description

In this study, we have used the dataset (https://www.kaggle.com/code/siddharthtyagi/stock-sentiment-analysis-using-news-headlines/input) that incorporates both global news and stock price changes from 2000 to 2016. The news headlines are scrapped based on the date and contain the labels for each day as 0 or 1. If there is a stock price increment at the end of the day then it will be represented as class 1 and the decrement is represented as class 0. In our work, based on the top 25 headlines, we predict the stock movement as either an increment or an decrement / the same stock price. The dataset contains 4101 rows and 27 columns. Of these columns, the 25 columns represent the top 25 news headlines, while the remaining two columns include the date of the headlines and the label indicating stock price movement. Specifically, the top 25 news headlines per day are included to represent worldwide events and trends.

3.2 Text Preprocessing

Text preprocessing is crucial in natural language processing (NLP) problems, particularly when dealing with textual data for tasks such as stock price prediction. In our approach, we implemented a comprehensive preprocessing pipeline to ensure the quality and relevance of the textual features extracted from news articles. The preprocessing pipeline consists of several key steps aimed at cleaning and standardizing the text data. Firstly, we removed punctuations and white spaces to ensure uniformity and readability of the text. Then, words containing digits were filtered out, as they often represent numerical data or noise rather than meaningful textual information in this context.

Stopwords, commonly occurring words with little semantic value such as "the," "is," and "and," were also removed to focus on content-rich terms. This step helps reduce the feature space dimensionality and improves the efficiency of subsequent analysis. Following the removal of stopwords, we converted all words into lowercase to ensure consistency in word representations. This step prevents the duplication of words due to capitalization variations and simplifies the subsequent processing steps. Stemming and lemmatization were then applied to further standardize the text data. Stemming represents reducing words to their root or base form by excluding suffixes, while lemmatization maps words to their canonical forms based on their meanings in the language. By applying these techniques, variations of words with the same root or meaning are consolidated, reducing redundancy and improving the accuracy of feature extraction.

3.3 Feature Extraction

After completing the preprocessing steps, we perform feature extraction using the Bag of Words (BoW) (HaCohen-Kerner et al., 2020), TF-IDF (Liu et al., 2022), and N-gram (Kim et al., 2021) models. The BoW model denotes each document's word frequencies as a vector, identifying the occurrence of each term in the corpus. Term Frequency-Inverse Document Frequency (TF-IDF) further enhances the Bag of Words representation by weighting the importance of terms using their frequency in the document and their rarity across the corpus. Finally, N-gram models capture the sequential dependencies between words by considering sequences of adjacent words as features. By implementing this comprehensive text preprocessing pipeline and leveraging advanced feature extraction methods including BoW, TF-IDF, and N-gram models, we ensure the quality and relevance of the textual features used for stock price prediction. This robust approach facilitates the extraction of meaningful insights from news articles and enhances the accuracy of predictive models in financial markets. Further, we performed normalization where it is crucial when dealing with features of varying units and scales in input data, as it scales the data between 0 and 1, aiding in quicker convergence. Utilizing the minmaxscaler function from the scikit-learn framework facilitates this process by computing the minimum and maximum values for each column.

3.4 Quantum Machine Learning Models

In this section, we trained various quantum machine learning models and compared the effectiveness of Quantum Logistic Regression (Ning et al., 2023), Quantum Support Vector Machine (Zhang et al., 2023), Quantum K-Nearest-Neighbour (Dang et al., 2018), Quantum Decision Tree (Beigi & Taghavi, 2020), Quantum variational classifier (Yano et al., 2021), and Quantum Random Forest (Shi et al., 2019) in predicting the movement of stock close price on a particular day based on the features extracted as discussed in Sect. 3.3. The models are trained with the news articles data and movement of stock price on specific days.

To apply the quantum machine learning algorithms, we converted the classical data into qubits using qiskit learn and pennylane libraries. Converting classical bits to qubits involves transforming information from a classical binary system (where bits can only be 0 or 1) to a quantum system (where qubits can exist in superpositions states of 0 and 1). The conversion process generally does not involve a straightforward one-to-one mapping, as qubits can hold much more information due to their quantum properties. The overview of conversion into qubits (Gandhudi et al., 2023) are listed as follows.

- Represent the Classical Bits: Let us consider a classical bit that can be either 0 or 1.
- Quantum Representation: In quantum computing, a qubit can exist in a superposition of states. So, we can represent a classical bit as a quantum state where both states (0 and 1) are present with certain probability amplitudes.
- Mathematical Representation: A qubit's state denoted as a linear combination of its basis states. So, a classical bit 0 denoted as $|0\rangle$ and a classical bit 1 denoted as $|1\rangle$. But in a qubit, we can have a superposition as $\alpha|0\rangle + \beta|1\rangle$, where α and β represents complex numbers, as probability amplitudes.
- Creating Superpositions: To represent classical bits as qubits, we typically create superpositions using quantum gates. For instance, the Hadamard gate can create a superposition of $|0\rangle$ and $|1\rangle$ with same probabilities.
- Encoding Information: We can encode classical information into qubits by manipulating these superpositions in various ways including quantum gates including Pauli-X gate or CNOT gate.

Quantum logistic regression (Gilyén et al., 2022) is a quantum Ml technique designed for binary classification problems, similar to its classical counterpart. Instead of using classical bits to represent data, quantum logistic regression employs qubits. Using quantum circuits the input data and parameters will be encoded, and quantum operations to perform computations and it aims to compute the optimal parameters that enhance the likelihood of observing the training data given a set of input features. Due to the quantum nature of the computation, it offers advantages in certain scenarios, such as processing large-scale quantum datasets.

Quantum SVM (Ding et al., 2021) is a quantum algorithm inspired by classical SVM, aimed at solving classification problems. Instead of depending on classical optimization techniques, it utilizes quantum algorithms for training and classification and it aims to compute a quantum-enhanced hyperplane that best segments different classes in the feature space, potentially offering improvements in computational efficiency and accuracy over classical SVM for certain problem instances.

Quantum KNN (Dang et al., 2018) is a quantum ML algorithm that uses the classical KNN approach. It operates on quantum data encoded in qubits and employs quantum operations to compute distances and classify new data points and it utilizes quantum interference to determine the class of a new data point based on its proximity to existing data points in the quantum feature space. While classical KNN can be computationally expensive for large datasets, quantum KNN may offer speedup for certain instances by exploiting quantum parallelism.

Quantum decision trees (Lu & Braunstein, 2014) are quantum counterparts to classical decision trees, designed for classification and regression tasks. They represent decisions using quantum states, accomplish computations using quantum gates and it leverages the quantum entanglement and superposition properties to efficiently explore and partition the feature space. By exploiting quantum parallelism and interference, they may offer benefits over classical decision trees in terms of computational speed and accuracy for certain datasets.

Quantum variational classifiers (Maheshwari et al., 2021) are quantum machine learning algorithms used for classification problems on quantum computers. They leverage variational quantum circuits and classical optimization techniques to classify input data and encode input data into quantum states, apply trainable quantum circuits to accomplish classification, and enhance the circuit parameters to optimize a cost function. They offer quantum advantage in classification problems by harnessing quantum properties including entanglement and superposition.

Quantum random forest (Srikumar et al., 2024) is a quantum version of the classical random forest algorithm, tailored for quantum computing environments. It comprises an ensemble of quantum decision trees trained on quantum data and it leverages the quantum computation principles to construct and combine quantum decision trees efficiently. By exploiting quantum parallelism and entanglement, it may offer improvements in computational efficiency and accuracy over classical random forests for certain problem instances.

4 Results and Discussions

Table 1 presents the results of various quantum machine learning classifier techniques applied to the stock price classification task. The classifiers include Quantum Logistic Regression (Ning et al., 2023), Quantum support vector machine (Zhang et al., 2023), Quantum K-Nearest-Neighbour (Dang et al., 2018), Quantum Decision Tree (Beigi & Taghavi, 2020), Quantum variational classifier (Yano et al., 2021), and Quantum Random Forest (Shi et al., 2019). The evaluation metrics used to compare the performance of the quantum classifiers are precision, recall, F1-score, and accuracy.

Based on Table 1, quantum random forest algorithm achieves a precision of 97%, surpassing other quantum classifiers. This signifies its ability to correctly identify a substantial portion of relevant samples. Moreover, it demonstrates a sensitivity of 94%, also outperforming other quantum classifiers. This suggests that a significant proportion of samples identified as relevant by the method are indeed relevant. Consequently, the value of accuracy stands at 96%, surpassing other quantum classifiers. Additionally, its F1-Score of 96% is the highest among all quantum classifiers, indicating a commendable balance between precision and accuracy.

Confusion matrices of the various quantum machine learning algorithms are presented as follows (see Fig. 2). Figure 2a represents the confusion matrix of Quantum Logistic Regression, Fig. 2b represents the confusion matrix of Quantum SVM, Fig. 2c represents the confusion matrix of Quantum KNN, Fig. 2d represents the confusion matrix of Quantum Decision Tree, Fig. 2e represents the confusion

Table 1 Comparison of classification values of quantum machine learning algorithms

Model	Precision	F1 - Score	Accuracy	False positive rate	Sensitivity	Specificity
Quantum logistic regression	0.93	0.82	0.84	0.05	0.73	0.94
Quantum SVM	0.85	0.74	0.77	0.10	0.65	0.89
Quantum KNN	0.89	0.84	0.85	0.08	0.79	0.91
Quantum decision tree	0.91	0.87	0.88	0.07	0.83	0.92
Quantum variational classifier	0.95	0.93	0.93	0.04	0.91	0.95
Quantum random forest	**0.97**	**0.96**	**0.96**	**0.02**	**0.94**	**0.97**

matrix of Quantum Variational Classifier and Fig. 2f represents the confusion matrix of Quantum Random Forest.

The ability to predict stock movements accurately bears significant managerial implications for both investors and organizations. For investors, precise forecasts enable informed decision-making regarding investment portfolios. This allows for optimized returns while mitigating risks. Additionally, strategic timing of trades based on accurate predictions empowers investors to seize market opportunities,

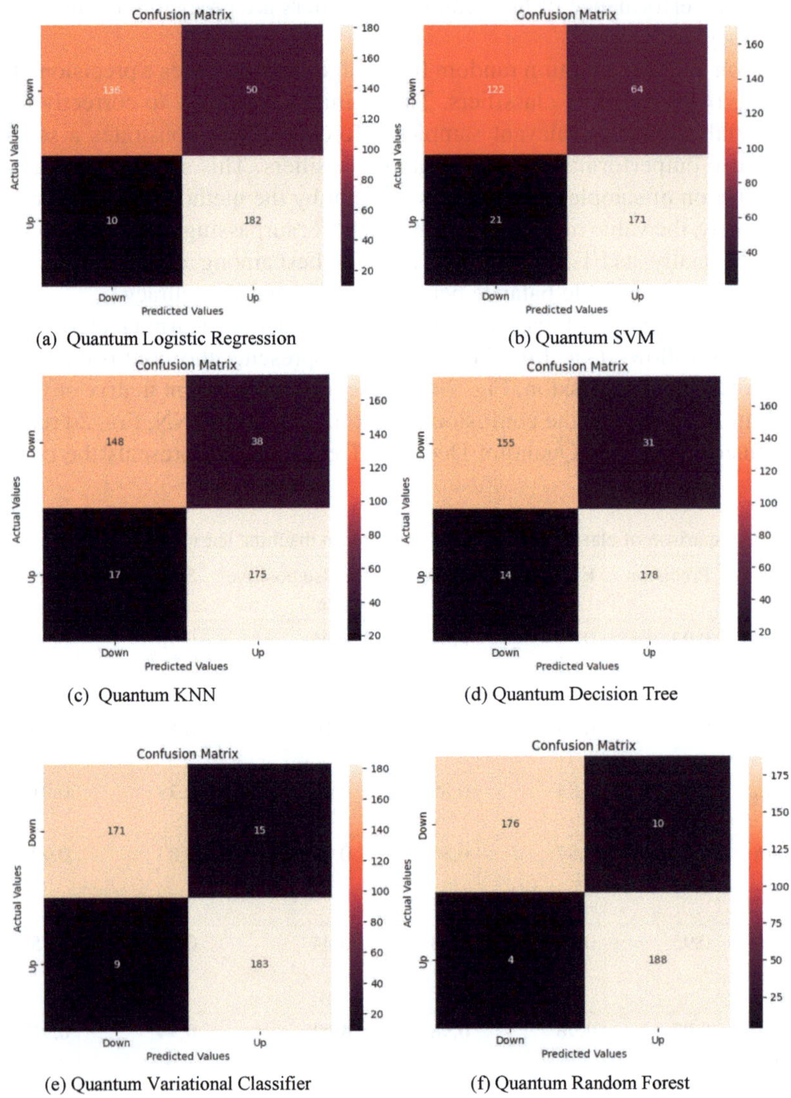

(a) Quantum Logistic Regression (b) Quantum SVM (c) Quantum KNN (d) Quantum Decision Tree (e) Quantum Variational Classifier (f) Quantum Random Forest

Fig. 2 Confusion matrices of various quantum machine learning algorithms (a) (b)

thereby maximizing profitability. For organizations, accurate stock predictions provide valuable insights for strategic planning. This facilitates better resource allocation, budgeting, and identification of growth prospects. Such informed decision-making not only enhances competitive advantage but also fosters investor confidence, potentially attracting increased investment and improving financial performance. Furthermore, accurate stock forecasts contribute to cost savings by helping to avoid poor investment decisions stemming from inaccurate predictions. Ultimately, the ability to predict stock movements with precision empowers both individuals and entities to navigate the financial markets complexities, confidently and effectively.

5 Conclusion and Future Work

Predicting stock price movements is significant in financial markets as it enables investors and organizations to perform informed decisions, manage risks effectively, and optimize investment strategies. In our work, efforts can be directed towards refining the quantum algorithms and models employed in the prediction process, aiming to improve accuracy and scalability. Quantum random forest algorithm emerges as a promising method in this regard, achieving an impressive sensitivity of 94% and precision of 97%. These metrics underscore its ability to accurately identify relevant stock movements, thereby enhancing decision-making processes. As financial markets continue to evolve, leveraging quantum computing approaches holds great promise for increasing decision-making processes and achieving superior investment outcomes. Applying and extending the quantum inspired machine learning capabilities would ultimately drive innovation and facilitate more accurate and reliable stock price predictions in dynamic market environments.

References

Albahli, S., Nazir, T., Nawaz, M., & Irtaza, A. (2023). An improved DenseNet model for prediction of stock market using stock technical indicators. *Expert Systems with Applications, 232*, Article 120903.

Ashtiani, M. N., & Raahemi, B. (2023). News-based intelligent prediction of financial markets using text mining and machine learning: A systematic literature review. *Expert Systems with Applications, 217*, Article 119509.

Bansal, M., Goyal, A., & Choudhary, A. (2022). A comparative analysis of K-nearest neighbor, genetic, support vector machine, decision tree, and long short term memory algorithms in machine learning. *Decision Analytics Journal, 3*, Article 100071.

Beigi, S., & Taghavi, L. (2020). Quantum speedup based on classical decision trees. *Quantum, 4*, 241.

Cho, J., Singh, M., & Lo, A. W. (2024). How does news affect biopharma stock prices?: An event study. *PLoS ONE, 19*(1), Article e0296927.

Choi, J., Yoo, S., Zhou, X., & Kim, Y. (2023). Hybrid information mixing module for stock movement prediction. *IEEE Access, 11*, 28781–28790.

Dang, Y., Jiang, N., Hu, H., Ji, Z., & Zhang, W. (2018). Image classification based on quantum K-Nearest-Neighbor algorithm. *Quantum Information Processing, 17*, 1–18.

Das, N., Sadhukhan, B., Bhakta, S. S., & Chakrabarti, S. (2024). Integrating EEMD and ensemble CNN with X (Twitter) sentiment for enhanced stock price predictions. *Social Network Analysis and Mining, 14*(1), 1–18.

Didugu, G., Gandhudi, M., Alphonse, P. J. A., & Gangadharan, G. R. (2024). VWFTS-PSO: a novel method for time series forecasting using variational weighted fuzzy time series and particle swarm optimization. *International Journal of General Systems*, 1–20.

Ding, C., Bao, T. Y., & Huang, H. L. (2021). Quantum-inspired support vector machine. *IEEE Transactions on Neural Networks and Learning Systems, 33*(12), 7210–7222.

Gandhudi, M., Gangadharan, G. R., Alphonse, P. J. A., Velayudham, V., & Nagineni, L. (2023). Causal aware parameterized quantum stochastic gradient descent for analyzing marketing advertisements and sales forecasting. *Information Processing & Management, 60*(5), Article 103473.

Gandhudi, M., Alphonse, P. J. A., Fiore, U., & Gangadharan, G. R. (2024a). Explainable hybrid quantum neural networks for analyzing the influence of tweets on stock price prediction. *Computers and Electrical Engineering, 118*, Article 109302.

Gandhudi, M., Alphonse, P. J. A., Velayudham, V., Nagineni, L., & Gangadharan, G. R. (2024b). Explainable causal variational autoencoders based equivariant graph neural networks for analyzing the consumer purchase behavior in E-commerce. *Engineering Applications of Artificial Intelligence, 136*, Article 108988.

Gilyén, A., Song, Z., & Tang, E. (2022). An improved quantum-inspired algorithm for linear regression. *Quantum, 6*, 754.

HaCohen-Kerner, Y., Miller, D., & Yigal, Y. (2020). The influence of preprocessing on text classification using a bag-of-words representation. *PLoS ONE, 15*(5), Article e0232525.

Hao, P. Y., Kung, C. F., Chang, C. Y., & Ou, J. B. (2021). Predicting stock price trends based on financial news articles and using a novel twin support vector machine with fuzzy hyperplane. *Applied Soft Computing, 98*, Article 106806.

Hu, L., Yu, Q., Jana, C., Simic, V., & Bin-Mohsin, B. (2024). *An intuitionistic fuzzy SWARA-AROMAN decision-making framework for sports event management*. IEEE Access.

Khan, W., Ghazanfar, M. A., Azam, M. A., Karami, A., Alyoubi, K. H., & Alfakeeh, A. S. (2022). Stock market prediction using machine learning classifiers and social media news. *Journal of Ambient Intelligence and Humanized Computing,13*(3), 1–24.

Kim, K., Gopi, S., Kulkarni, J., & Yekhanin, S. (2021). Differentially private n-gram extraction. *Advances in Neural Information Processing Systems, 34*, 5102–5111.

Kobayashi, N., Suimon, Y., Miyamoto, K., & Mitarai, K. (2023). The cross-sectional stock return predictions via quantum neural network and tensor network. *Quantum Machine Intelligence, 5*(2), 46.

Koratamaddi, P., Wadhwani, K., Gupta, M., & Sanjeevi, S. G. (2021). Market sentiment-aware deep reinforcement learning approach for stock portfolio allocation. *Engineering Science and Technology, an International Journal, 24*(4), 848–859.

Kumbure, M. M., Lohrmann, C., Luukka, P., & Porras, J. (2022). Machine learning techniques and data for stock market forecasting: A literature review. *Expert Systems with Applications, 197*, Article 116659.

Li, Y., & Pan, Y. (2022). A novel ensemble deep learning model for stock prediction based on stock prices and news. *International Journal of Data Science and Analytics, 13*(2), 139–149.

Li, Q., Tan, J., Wang, J., & Chen, H. (2020). A multimodal event-driven LSTM model for stock prediction using online news. *IEEE Transactions on Knowledge and Data Engineering, 33*(10), 3323–3337.

Li, Y., Wang, Z., Han, R., Shi, S., Li, J., Shang, R., Zheng, H., Zhong, G., & Gu, Y. (2023). Quantum recurrent neural networks for sequential learning. *Neural Networks, 166*, 148–161.

Liu, Z., Tseng, H. K., Wu, J. S., & Ding, Z. (2020). Implied volatility relationships between crude oil and the US stock markets: Dynamic correlation and spillover effects. *Resources Policy, 66*, Article 101637.
Liu, H., Chen, X., & Liu, X. (2022). A study of the application of weight distributing method combining sentiment dictionary and TF-IDF for text sentiment analysis. *IEEE Access, 10*, 32280–32289.
Lu, S., & Braunstein, S. L. (2014). Quantum decision tree classifier. *Quantum Information Processing, 13*(3), 757–770.
Lu, M., & Xu, X. (2024). TRNN: An efficient time-series recurrent neural network for stock price prediction. *Information Sciences, 657*, Article 119951.
Ma, D., Yuan, D., Huang, M., & Dong, L. (2024). VGC-GAN: A multi-graph convolution adversarial network for stock price prediction. *Expert Systems with Applications, 236*, Article 121204.
Maheshwari, D., Sierra-Sosa, D., & Garcia-Zapirain, B. (2021). Variational quantum classifier for binary classification: Real vs synthetic dataset. *IEEE Access, 10*, 3705–3715.
Maqbool, J., Aggarwal, P., Kaur, R., Mittal, A., & Ganaie, I. A. (2023). Stock prediction by integrating sentiment scores of financial news and MLP-regressor: A machine learning approach. *Procedia Computer Science, 218*, 1067–1078.
Moews, B., & Ibikunle, G. (2020). Predictive intraday correlations in stable and volatile market environments: Evidence from deep learning. *Physica a: Statistical Mechanics and Its Applications, 547*, Article 124392.
Mu, G., Gao, N., Wang, Y. and Dai, L., 2023. A stock price prediction model based on investor sentiment and optimized deep learning. *IEEE Access*.
Nekrasov, A., Teoh, S. H., & Wu, S. (2022). Visuals and attention to earnings news on Twitter. *Review of Accounting Studies, 27*(4), 1233–1275.
Ning, T., Yang, Y., & Du, Z. (2023). Quantum kernel logistic regression based Newton method. *Physica a: Statistical Mechanics and Its Applications, 611*, Article 128454.
Ray, P., Ganguli, B., & Chakrabarti, A. (2021). A hybrid approach of bayesian structural time series with lstm to identify the influence of news sentiment on short-term forecasting of stock price. *IEEE Transactions on Computational Social Systems, 8*(5), 1153–1162.
Shi, Y., Wong, W. K., Goldin, J. G., Brown, M. S., & Kim, G. H. J. (2019). Prediction of progression in idiopathic pulmonary fibrosis using CT scans at baseline: A quantum particle swarm optimization-Random forest approach. *Artificial Intelligence in Medicine, 100*, Article 101709.
Srikumar, M., Hill, C. D., & Hollenberg, L. C. (2024). A kernel-based quantum random forest for improved classification. *Quantum Machine Intelligence, 6*(1), 1–31.
Swathi, T., Kasiviswanath, N., & Rao, A. A. (2022). An optimal deep learning-based LSTM for stock price prediction using twitter sentiment analysis. *Applied Intelligence, 52*(12), 13675–13688.
Valle-Cruz, D., Fernandez-Cortez, V., López-Chau, A., & Sandoval-Almazán, R. (2022). Does twitter affect stock market decisions? financial sentiment analysis during pandemics: A comparative study of the h1n1 and the covid-19 periods. *Cognitive Computation, 14*(1), 372–387.
Wan, X., Yang, J., Marinov, S., Calliess, J. P., Zohren, S., & Dong, X. (2021). Sentiment correlation in financial news networks and associated market movements. *Scientific Reports, 11*(1), 3062.
Wang, Z., Hu, Z., Li, F., Ho, S. B., & Cambria, E. (2023). Learning-based stock trending prediction by incorporating technical indicators and social media sentiment. *Cognitive Computation, 15*(3), 1092–1102.
Wen, D., Zhao, T., Fang, L., Zhang, C., & Li, X. (2024). MWDINet: A multilevel wavelet decomposition interaction network for stock price prediction. *Expert Systems with Applications, 238*, Article 122091.
Xie, Q., Fang, T., Rong, X., & Xu, X. (2024). Nonlinear behavior of tail risk resonance and early warning: Insight from global energy stock markets. *International Review of Financial Analysis*, 103162.
Xu, Z., Wang, Y., Feng, X., Wang, Y., Li, Y., & Lin, H. (2024). Quantum-enhanced forecasting: Leveraging quantum gramian angular field and CNNs for stock return predictions. *Finance Research Letters, 67*, Article 105840.

Yano, H., Suzuki, Y., Itoh, K. M., Raymond, R., & Yamamoto, N. (2021). Efficient discrete feature encoding for variational quantum classifier. *IEEE Transactions on Quantum Engineering, 2*, 1–14.

Zhang, R., Wang, J., Jiang, N., & Wang, Z. (2023). Quantum support vector machine without iteration. *Information Sciences, 635*, 25–41.

Zhao, Y., & Yang, G. (2023). Deep Learning-based Integrated Framework for stock price movement prediction. *Applied Soft Computing, 133*, Article 109921.

Data Analytics on Resilience of Mutual Funds Through Sustainable Investments and Its Impact on Volatility of Returns for Decision Making Towards Business Excellence

K. Vinitha and P. Saravanan

Abstract *Purpose* Companies having good "Environmental, Social, and Governance" (ESG) score do not have to face any sort of lawful and reputational risk which can create negative effect on financial performance of the companies thereby they could help investors to evade the company which are at risk. Through this study the impact of sustainability score in Mutual fund's return volatility is analysed which would be helpful for the Investors. A company having good ESG score may bring in the investors as the company is being protected from the forthcoming risks such as pollution, low governance and so on. *Design/Methodology/Approach*—The overall effect of Mutual funds of companies based on sustainability rating is analysed in this paper. A sample of 64 Mutual Funds and the rating given by Morningstar in the year April 2023 is examined. The period of research is for one year, the researcher through this paper, analysed the situation whether the mutual fund companies are resilient to the COVID crisis which the economy faced. A cross sectional regression analysis is done with the data, as cross-sectional regression is primarily meant for an event-based study basically for one year. *Findings*—The outcome indicated that the Mutual Funds of companies with social risk score, governance risk score and higher sustainability score depicted high return volatility. The result could be beneficial for investors, business, and stakeholder at large. A high return volatile market implies the potential ability of stocks to yield well in the stock market on a long run. *Research limitations/ implications*—This research focussed on mutual funds by analysing the ESG scores and the return volatility of Mutual funds. In this study explanatory research is done as it delivers added information about the observed object and its relation with the environment. This paper focussed on Mutual funds of 64 companies due to limitations of data of few companies. *Practical Implications*—The sustainable investments are on the trajectory of resilience and ESG scores has proven to be lens which can aid business leaders to come up with greater number of novel ventures. The findings can

K. Vinitha (✉) · P. Saravanan
SRM Institute of Science and Technology, Kattankulathur, Chennai, India
e-mail: vinithak@srmist.edu.in

P. Saravanan
e-mail: saravanp2@srmist.edu.in

B. Basu et al. (eds.), *Data Analytics for Decision Making towards Business Excellence*, Uncertainty and Operations Research, https://doi.org/10.1007/978-981-95-2429-7_8

be considered by investors to invest in high ESG score companies which could gain investors with high return in terms of both value and profit, whereas it also paves the way for the companies with low ESG scores to consider the trajectory to raise the ESG scores in terms of the resources where ever lagging behind. *Social Implications*—It is high time for the Mutual fund companies to take into consideration that what they gain from the society as profits whether they are returning it back to the society in terms of ethical practices which need to be followed. It is unethical, if the companies are not taking any steps to gain sustainability even after much deliberations on climate change and sustainable goals. Through this paper the mutual fund companies with low ESG score are walking on thin ice. *Originality/Value*—There are lot of research studies available on Exchange traded funds and their ESG impacts. This research is the pioneer to expand the study in order to find out the impact of mutual fund performance towards sustainable Investments. And this paper comes out with an event-based study as it tells whether Mutual fund companies are resilient enough immediately during the post COVID time period, which in turn is pioneer research.

Keywords Sustainable investments · Data analytics · Business excellence · MF investment decisions · ESG Risks · Mutual Funds · Cross-sectional

1 Introduction

Adversities have brought significance for the need of taking actions to safeguard the business and society from long tenure sustainable threat. Crisis quickens nascent ESG trends, and the enhanced concentration on company's environmental and social impact may ever last even after the happening of any crisis. Three fundamental concepts that guided on sustainable investments on the basis of environment, societal and governance factor are taken basically for analysis on ESG scores. It takes us to a sound investment rather than an individual investment. When ESG variables are taken into consideration focus should be given on long-term returns. Investments are based on the goals and interests of the investors, which may give the investors a reply to the yes or no question. Focus is also given to short-term and long term returns to achieve the objective. This research will help businesses to attract and obtain finance which may also assist the leaders to clarify the effect of ESG on the investment criteria of an organization. Research was taken by the CFA society which has taken the comparison of the recommendation put forward by SEBI's consultation paper which had the international regulation such as the SFDR and voluntary initiative such as the CFA Institute's ESG Disclosure Standard for Investment product and came up with the recommendation which aligned in all respect.

A most promising challenge for fund manager is the difficulty to demonstrate the actual world the effect and the outcome of the ESG product. Almost all funds take ESG risk and opportunity which they create investment portfolio, but they don't take the effect of portfolio company on environment and society. Even though they may be interlinked, fund manager may find it easier to report on the former than on the latter. Moreover SEBI has acknowledged this kind of distinction in the recent paper on ESG rating, where a variance among ESG risk rating and impact rating are seen. ESG risk ratings dealt with the effect of social or environmental issue on the company value and the impact rating deal with effect in the next direction. Secondly the challenged aspect by the participant is the probable duplicate of disclosure, especially around stewardship reports.

Over the previous years, sustainable investments were able to catch the eye of investors on a world-wide, as people look for investments on and on. Sustainable investments which are also called as impact investments otherwise socially responsible investments. This method of investment takes into consideration both finance return and environment-oriented return, and it also incorporate social and corporate governance objective. Environment feature targets the carbon fuel, the social feature focus on work environment characteristics such as tobacco avoidance, weapon produce so on. The goal is to bring in competing monetary results constantly; and to give a helping hand for companies to come up with a stakeholder-oriented archetypal of commercial comportment in a long-standing basis's Investments are more demanded in current scenario in India as well as in other countries due to the surge of climate change inferences and the societal consequence caused due to Covid-19.

Vidhu Shekhar, of CFA Institute, found that the world-wide sustainable investment is found to be in a robust up surge and attained $35 trillion or 36% of the overall assets in management in 2020. India had shown that every sustainable investment product which has been plunged in the previous 2 years, had highlighted incredible surge potential. ESG fund are showing up surge in India in mutual fund sector as asset management company has launched equity schemes in the ESG space. They also have launched exchange traded funds. Currently it has been reported to have twelve MFs in India which have "ESG" as target and they consider the insights of the stakeholders. SEBI projected that "Credit Rating Agencies and Research Analysts (CRARA)", who devise a slightest net worth of ₹ 10,00,00,000/- have the eligibility for accreditation for ESG rating.

Socially and Responsibly (SRI) conscious Investment (SRI) substitutes are introduced to traditional risk-return financial market techniques in order to acknowledge wider stakeholder pressures on corporations. In addition to extra latest community gravities to guard alongside short-sighted behaviour, institutional investors and regulatory stakeholders have become advocates for SRI. Over the last twenty years, socially responsible investing (SRI) has evolved from being the purview of a minor digit of specialized trade venture moneys to being speculation ethos embraced through an increasing number of big investment establishments, such as insurance firms and huge pension funds. Investors can align their investment policies with their moral and ethical principles by using this unconventional investment strategy.If profits from upper yields and/or inferior risks may also be realized via "SRI" portfolios that outperform traditional investing portfolios in terms of returns and hazards, then "SRI" may also offer extra potential benefits to investors. On the other hand, little is known regarding the systematic accumulation of these benefits.

1.1 Current State of Indian Mutual Fund Industry

According to AAUM [Average Assets under Management], the mutual fund industry in India is worth of ₹ 61,337,227 crore as of June 2024. The assets under management, or AUM, are valued at ₹ 61,15,582 crore. This is a six-fold gain over the previous ten years, from ₹ 9.75 trillion in June 2014 to ₹ 61.16 trillion in June 2024. Such remarkable rise, which reflects strong market performance and rising investor confidence, demonstrates the mutual funds' growing appeal and importance among Indian investors. The industry's crucial significance in the financial environment and its bright future trajectory are highlighted by the notable increase in AUM.

1.2 AUM Growth in the Indian Mutual Fund Sector Over the Previous Five Years

Even though the global financial markets have slowed down over the last five years due to the COVID-19 pandemic, as well as the turmoil and wars in Europe and the Middle East, the Indian mutual fund industry has proven incredibly resilient and has grown significantly. In the last five years, the mutual fund industry's assets under administration have doubled, rising from ₹ 22.6 lakh crore in 2019–2020 to ₹ 54.1 lakh crore in 2023–2024. (www.bajaj.finserv.in).

1.3 ESG Ratings' Effects on Investments

ESG criteria have shown themselves to be extremely helpful as investors continue to include ESG scores into their strategies; ESG-focused portfolios routinely outperform standard portfolios. According to a Frontiers study, trades that contain "ESG" impressions into their strategies classically pull in a wider group of savers that place a higher priority on societal accountability and defensible. Better fiscal enactment and more principal pour might effect from this. In a similar vein, a meta-analysis of more than 2000 studies has verified the verifiable economic and ethical rationale for ESG investment (simply sustainable.com).

1.4 Why is a Strong ESG Framework Essential to Promoting Sustainability in the Future?

India is the seventh most vulnerable nation to climate change, with droughts, heat waves, and floods impeding GDP growth, according to the Global Climate Risk Index 2024. The nation lost billions of dollars in the manufacturing, construction, service, and agricultural sectors. Strong adaptation and mitigation plans are ensured by incorporating ESG concepts.

The EU receives 6% of India's exports of iron ore pellets, steel, and aluminum products. CBAM (Carbon Border Adjustment Mechanism) will affect these products. For India, not implementing ESG would be a major worry. Beginning on January 1, 2026, the CBAM will apply to seven carbon-intensive industries. In order to realize India's ambitious goal of becoming a $5 trillion economy, sustained growth is essential (alp.consulting).

There are still unanswered questions about the possible systematic benefits that SRI portfolios could offer stockholders in the event of a monetarist catastrophe like the universal monetary crunch of 2008. Fund managers and other investors are aware of the benefits of placing money into portfolios that see a quick rebound in value following a pecuniary shockwave. Investors then need to consider how resilient their portfolios are to these kinds of shocks and whether SRI portfolios are more resilient than traditional portfolios. The literature on sustainability and finance acknowledges that resilience has two components: inherent and adaptive. The majority of financial resilience research focuses on the quick, short-range response to a disaster that results in value rescue. Inherent resilience is related to how crises are handled by way of part of industry as usual at a certain point in time. Adaptive resilience, a dynamic concept linked to the rate of recovery from a disaster to a desirable state, like the gradual refurbishment of assortment worth, is of higher importance for this study. After a crisis, bond holder favour assets with quicker rates of value recovery, indicating that further research is needed to develop instruments for financial markets' adaptive pliability. Depositors will compare the results from conservative ranges with those from SRI portfolios in order to incentivize companies to adopt sustainability practices. Investors should consider if traditional (modern) or "SRI portfolio investments" offer better overall benefit, particularly when it comes to their ability to adapt and endure during difficult times. Therefore, the purpose of this research is to examine whether investors in SRI portfolios systematically acquire benefits in terms of both risk and return, as well as if these benefits persist during financial crises. Therefore, the study poses the following question: Do risk-averse investors benefit more from equity investments in portfolios made up of sustainable and ethical businesses than they do from traditional portfolios in footings of adaptive resilience? The following two questions will determine this: Do comparative benefits for investing in "SRI portfolios" systematically accumulate ended period? Furthermore, In times of financial crisis, do the relative benefits of investing in SRI portfolios hold true?

The research work is prearranged into eight sections. They are "Objectives of the study", " The theoretical framework", "Literature review and hypothesis development", "Describes research design and methods", " Data gathering and Analysis"," Findings and Discussion" examines Mutual Fund's volatile by means of CSRA; and "Conclusion of the study and Future Research" and "References".

2 Objective of the Study

Explored that assets under management (AUM) for socially responsible investment (SRI) mutual funds witnessed an upsurge of more than 13% from 2007 to 2009, on the other hand conventional funds identified meagre variations. Benson and Humphrey (2008) supported with proof the idea that SRI funds depicted high stability in cash flows, and Bollen et al. (2007) found that SRI mutual funds tend to be more responsible to positive returns and less affected by negative ones. According to Dolvin et al. (2017) Morningstar Sustainability metrics did not show any impact on fund flows in the US market. Anyhow, for funds which identified as SRI, these metrics showed a positive influence, which eventually made them more alluring to investors. El Ghoul and Karoui (2017) explored that funds with high corporate social responsibility (CSR) are likely to attract investors who are less sensitive to performance, although the effect of CSR on flow persistence is statistically insignificant. Consistent with prior research, we posit that funds with higher environmental, social, and governance (ESG) scores attract meticulous investors who line up values than performance, which lead to low sensitiveness in fund flows to past performance. Hence there is a need to study the impact of sustainability score in volatility of Mutual Fund.

Sustainable Mutual fund portfolios are not stable with advanced Portfolio Theory, in which the theme is investment in various asset classes. Other argument states, "sustainable funds" devour limit on their asset special, so there is higher menace due to the minimal diversified assets. Previous literature states that risk adjusted performance metrices have been utilised of measure monetary effect of ESG, but does it had an effect on lower side risk was not found completely. Today management of risk is one among the most challengeable task for the firm, and their attention is towards risk and sustenance for avoidance of risk. Haitham Nobanee (2020) focused on risk and sustenance for avoidance of risk, especially in financial management. For attaining this objective, a methodology on secondary data was taken where 25 articles were taken from Scopus and five from SSRN. The outcome revealed the significance of risk management and sustenance for organization. But in these papers the risk and sustenance were not taken together for studies. Hence new methods were developed for avoidance of risk to have sustenance. "Olitzky & Benjamin's (2001)" meta-syntheses explored significance of adverse connexion among business societal perform metrics and risk and, the association is strong for markets-basis measure of risks against secretarial- oriented measure. outcome indicated how effective the tool was, in order to identify and analyse sustainable risk and to derive and monitor strategy thereby in order to lead to an increase in awareness of the connections among societal, ecological and sustainable aspect and business implication. Hence it renders backing for company in the ways to work with developing the product so that it can contribute to transition of society for sustainability, in a way it benefits the organization.

Joris-Johann Lenssen et al. (2014) described the significance to have a united approach to report on sustainable risk, with an answer at all levels including individual wise, sector wise and in governance. Hence the research contribution advanced to governance in order to minimise risk in bringing in business sustainability. So another objective of this paper is to find out whether the "ESG score" positively associated with return volatility of Mutual Funds.

3 Theoretical Framework

Prevailing literatures present different viewpoints underwriting to the continuing discussion on the enactment of funds integrating communally accountable principles and their arrangement through prosperity intensification purposes. The relationship amid "ESG" and financial performance can be clarified by 2 disparate theoretic belvederes. Certain readings emphasize the possible adverse influence of "ESG performance" on "MF" performance. Spending in firms thru extraordinary "ESG scores" can lead to a decrease in deal chances and an rise in observing costs. The underperformance of SRI funds compared to conventional counterparts can be attributed to their exclusion of investments that do not align with "ESG" objectives, thus missing potential financial opportunities. Essentially, devoting based on "ESG" criteria limits investment choices as socially responsible investments constitute a subset of the overall universe of investment opportunities. This implies that ethical investors may need to forego some diversification benefits. According to portfolio theory, introducing additional constraints, such as ESG criteria, into the asset selection process is likely to impede optimal portfolio construction. Specifically, these constraints reduce the pool of available investments, affecting potential diversification compared to unconstrained portfolios. Altered diversification benefits may result in lower returns for constrained portfolios relative to unconstrained portfolios. Additionally, the need for screenings and rankings to assess the level of investment social responsibility incurs costs that can impact the net returns of investments.

On the other hand, arguments in favor of socially responsible investments highlight the long-term benefits, such as improvements in economic performance, that can be derived from investing in firms that follow a strategy enhancing corporate social responsibility. A higher level of corporate social responsibility reflects a higher quality of corporate management, potentially leading to comparative advantages relative to firms with lower levels of corporate social responsibility. Thus, mutual funds investing in stocks with high "ESG" performance can generate higher levels of performance. According to this perspective, the additional costs due to ESG screenings may be offset by excluding unprofitable firms through the screening process. The outperformance of SRI funds compared to their conventional peers can be explained by the fact that fund managers seeking firms with high ESG scores are likely to incorporate those with strong financial fundamentals or sustainable performance into their portfolios, ultimately resulting in better performance. The theory argues that a firm should create value for all stakeholders, not just shareholders. In 1984, R. Edward Freeman originally detailed the Freeman came up with stakeholders' theory there by a theory-oriented basis for ideas which included Corporate Social

Responsibility (CSR), socially responsible investments, combined reports, sustainable investments and ESG reports. With regard to the stake holder theory, CSR paved the approach aimed at organizations' prolonged monetary gain, as it is more towards building a trust among worker community, customer relationship, society at large, environmentalists and so on. An organization which has connectivity with lots of elements, which is the, stakeholder who had or can claim to be owner, right, or interest in a corporate and its activism, whether it is in pastime, current, or may be about to come (Clarkson, 1995, page 106), that can be proven to be the significant factor to a corporate's monetary affairs to perform. The stakeholder theory tells that poor ESG performance move to get higher incentive for sustainability disclosure for enhancing legitimacy and to direct an action to stakeholder which organizations were really good than actually professed.

4 Review of Literature & Developing Hypothesis

A crucial component of VaR computation is volatility estimation, which is based on the time horizon, confidence interval, and predicted volatility of the continuous returns being studied. VaRmodeling is a logical use of volatility estimates for ex-post volatility predictions, even if VaR is a measure of the tail losses. VaR's performance in volatility projections and its bounds versus ex-post benchmarks have been examined in a number of research. Ex-post volatility is typically underestimated by VaR. According to Brooks and Persand (2003), VaR forecasts are better when the volatility criteria are simpler. According to, less complex VaR models frequently result in an underestimating of the VaR. For instance, when fat tails are not taken into consideration, the widely used VaR methodology, which is established a normalcy notion on the conditional dissemination of returns, frequently results in an underestimating of financial losses.

Unexpected losses may occur for investors and financial institutions if actual returns diverge from the VaR projection. This has prompted researchers to use back testing techniques to find model specifications that best capture the instability subtleties of asset proceeds & assumptions that yield best overall volatility forecasts. Key specifications influence measure's ability to produce accurate forecasts are discussed. These include estimated factors for the asset return distribution, the prototypical for the conditional volatility distribution's tail behavior, and the various methods based on daily or intradaily data. The above-mentioned brief examination of the literature demonstrates that opinions on the viability of accurate extrapolative (before) estimations of the (after) instability of monetary chattels are divided. Drawing from the widely recognized proverb commonly ascribed to statistician George Box, "Every model is incorrect, but some are valuable," examined the collected works to evaluate the VaR technique for estimating ex-ante unpredictability and contrasting it with ex-post realized volatility.

Haider Mohammed Baniatal (2024) The persistence of this article is to assess the return concert of conventional mutual funds (CMF) and Islamic mutual funds

(IMF) during two stretches of crisis: the "COVID-19 pandemic" crisis and the global financial crisis (GFC). This research offers distinctive proof of how a global infectious disease affects mutual funds. The three crises from January 2000 to October 2020 are covered by the 87 equity IMFs and 83 equity CMFs in Malaysia that make up the entire sample. The one-factor and four-factor models are the two market risk-adjusted performance metrics used in this study to determine the overall performance of the funds during the crisis. The primary findings indicate that, on average, during the course of the sample period, both IMFs and CMFs underperform the market return. The crises had an impact on both IMFs and CMFs, but IMFs performed better during the crisis than CMFs.

Pablo Vilas et al. (2024) observed through a Machine Learning (ML) approach as "past ESG performance guarantees future ESG performance," which could be of curiosity to publically conscious savers, in toting to verifying that "past financial return does not guarantee future financial return."

Sara Ali et al. (2024) contend that the information provided by a fund's ESG rating can enhance the Flow- performance association between high "ESG-rated (or sustainable) funds" by mitigating certain of the informational irregularities that might exist among investors and stock bosses regarding their adherence to ethical business practices.

Alexandre Momparler et al., (2024) by using ML models have shown how important ESG ratings are in the intricate world of predicting mutual fund success. The ESG score confirmed its position as a critical component of the contemporary investment environment by ranking among the top five most significant criteria in forecasting mutual fund performance. This research shows that, without sacrificing financial results, investors can match their financial goals with their dedication to sustainability and ethical investing.

Fund flows managed by banks or businesses in the same financial group as banks are less erratic and less susceptible to poor historical performance. Compared to their unaffiliated counterparts, this allows bank-affiliated funds to keep smaller precautionary cash buffers and weather hardships better. Banks purchase shares in distressed connected funds that are seeing significant outflows in order to support their liquidity. As a result, investors' redemptions are less severe due to strategic complementarities. The parent company's financial stability is a prerequisite for receiving liquidity support and other advantages of bank affiliation. Through ownership linkages, the mutual fund industry is impacted by the banking system's distress.

1970 was a time where numerous CEOs showed favouritism towards Fried man's doctrine – which was about the significance of shareholder. Friedman (1970) argued about execution of corporate social responsible principle which imposed cost on shareholder, which showed to be baseless, not only that it has also impacted unfair tax on shareholder. Bowen (1953) suggested, corporate resource to be utilised to meet social needs, corporate charity and volunteer activism; business have to take decision with the base on the value of society. ESG reporting states that it could deliver beneficial informations to investor, supplier, employee, customer, regulator, society and so on (Al Amosh & Mansor, 2021). So maintainable investment may outclass, and it is explicit for the good sphere anywhere firm "do well by doing good". Mathews

(1997) argued, sustainable report has to be looked as 'socially contractual' firm which utilise sustainable reports for surging legality with a view to operate in such a way which is respectable to the humanity. suggested, firm which have low ability to perform in the environment as per the media show likeliness to pursue legality through sustainable disclosure. In the recent days societal concerns and environment concerns grabbed more focus. Manes Rossi et al. (2018) explored, company show a general cognizance towards societal, environment revelation for maintaining legitimate and in order to come up with investor and stakeholder expectations. Active firms volunteer and give extra ESG inputs in annual reports for giving an image of incorporating well doing and operating criteria have undergone to go beyond stakeholders' estimations. Dyllick and Hockerts (2002) suggested that corporates socially responsible report conveyed significant inputs to stakeholder as to there might be business activities which can sustain in the maintenance of environment, societal and economic targets. Brammer and Pavelin (2006) argued, corporates revelation of sustainable reports doesn't denote corporate's commitments in the reduction of environmental aspects nor societal risk, in its place the goal to disclose is to impact stakeholder perceptions with regard to firm's forthcoming monetary probabilities.

Topping (2012) suggested, ESG indicator building enterprise activity in incorporating corporate socially responsible aspects which is assessable and explicit to external stakeholders. "Al Amosh and Mansor (2021)" suggested, combined reports enhance the informational quality and the move for a world-wide acceptance of combined reports which in fact reflect the expectation of stake holder. Firms which do not do well in environment and societal aspect and doesn't have disclosed, might be subjected to contrary publicities, awareness raising, or boycotting camps by consumers and stakeholders group.

Hence the succeeding hypotheses are formulated:

H1: "The sustainability score is positively associated with volatility in returns of Mutual Funds."

H2: "The Environmental, Social and Governance Score are positively associated with volatility in returns of Mutual Funds".

5 Research Design and Methodology

We have gathered sustainability rating data from Morningstar, focusing on a sample period spanning from September 2022 to June 2023, comprising 64 observations. The sustainability rating is determined by the weighted mediocre sustainability notches of a fund's fundamental holdings, provided by Sustainalytics, which evaluates environmental, social, and governance (ESG) aspects. On a monthly basis, Morningstar calculates the weighted average sustainability scores to create sustainability score at the fund share-class level. Funds are then categorized founded on their sustainability scores into "High (5 globes)", "Above Average (4 globes)", "Average (3 globes)",

"Below Average (2 globes)", "Low (1 globe)". This globe rating system serves as an objective metric for investors to assess how investments align with ESG challenges. Our analysis primarily relies on the Morningstar Category variable, taking into account Sus score, Env score, Soc Score, Net worth, and excluding bonds, RTF, board reserves, and debt-based. Return volatility is estimated exhausting yields from the last eighteen months. This methodology is inspired by previous studies, such as "Hartzmark & Sussman (2019), Pastor and Vorsatz (2020), and Pastor, Stambaugh, and Taylor (2015)". The goal is to provide investors with a comprehensive understanding of how mutual funds align with various sustainability criteria. With the use of financial variables, we hope to develop a strong regression prediction model that can pinpoint the most important characteristics or critical elements that predict fund performance. This regression model's goal is to identify the characteristics of a fund that generates high returns while avoiding overfitting and producing forecasts that may be applied broadly. Part of our goal is to challenge ESG funds' professed social responsibility missions. It is challenging to determine with objectivity whether the companies that underpin these funds are in fact socially conscious. Previous research that looks into this issue depends on a social responsibility label that has been given by an outside, possibly reliable source "(such as the Domini Social Index, PRI principles, or vendor-provided ESG scores)". Studies make the implicit assumption that 3rd party endorsements or labels of social responsibility accurately reflect a company's governance, social, and/or environmental attitude. The validation of these evaluations is the subject of a growing body of scholarship. Berg (2022) study shows that there is little correlation between the ESG scores given to the same companies by various rating agencies, whereas additional "(Serafeim and Yoon, 2021)" emphasizes that the relationship between "ESG" bulletin values is weakened by ESG rating dispute. Additionally, recent research demonstrates that these ratings (i) act as though they include shares that are positively mentioned in the media but do not react to metrics of causal investor behaviour "(Yang, 2021)" & largely interrelated thru amount of revelation, and thus unconsciously per firm dimension, then not with the info presence unveiled "(Drempetic et al., 2017; Lopez-de-Silanes et al., 2019)."

In this study explanatory research is done as it delivers added information about the observed object and its interaction with the environment. Here in this paper the effect of sustainable scores on the monetary perform analytics of Mutual fund is observed by making use of a sample of 64 companies Mutual Funds, in which the ratings given by Morningstar in April 2023, is taken and "CSRA" is done to find out the impact of ESG scores on volatility and financial outcome based on Lan Sun and Garrick Small (2021) a study carried out in Australia. The current study is pioneer to research about Mutual funds with regard to ESG scores and its impact on Volatility scores and impact on financial outcomes in India. In this paper "CSRA" is done with Mutual Fund's monetary performance in association to sustainability during 2023 is examined. As this is a period where COVID -19 has hit the economy and where, the researcher needs to analyse the situation on the basis of the crisis.

The model consisted of Volatility as a function and assessed ESG pillars with Net Asset value as the control variable. Cross sectional OLS estimation is done to deliver its maximum outcome, White (1980) base paper reference to be given is adopted for estimation procedure to come up with heteroscedasticity robust standard errors.

$$\text{Volatility}_{\text{i}} == \text{a}_{\text{i}} + \beta_1\text{Envscore}_{\text{i}} + \beta_2\text{Socscore}_{\text{i}} + \beta_3\text{Net Assets}_{\text{i}} + \varepsilon_i \quad (1)$$

$$\text{Volatility}_{\text{i}} == \text{a}_{\text{i}} + \beta_1\text{Susscore}_{\text{i}} + \beta_3\text{Net Assets}_i + \varepsilon_i \quad (2)$$

Three years return, is taken as the compound annual growth rate for "Mutual Fund" for the previous 3 yrs. Here 3 years standard deviation is used to assess volatile index. 3 years standard deviation for MFs on the basis of month wise return for the previous3 years constructed on the "Regression of a fund's" month wise return on Standard & Poor five hundered reappearance. The sustainable levels "(SusScore)", a Morningstar's (MS) score range "0–100", assess the assortment to performi in relation to the level of sustainable indexes. Environment ("EnvScore"), Societal ("SocScore"), "Governance risk" are MS scores starting "0–100", assess portfolio to performi in relation to environment, societal and governance risk.

6 Data Gathering and Data Analysis

A sample of 64Mutual Funds rated by Morningstar in April 2023 is depicted below. Table 1 states the statistical indexes. The SusScore has a mean value of 29.6800 point, scores are brought to a normal value of "0–100" scale on a mean of fifty, the sustainable level of MF is some what lower than the average score. The "mean" for Envscore, SocScore and GovScore are 4.2640 points, 7.5930 points, and 7.0070 points. It implies Mutual funds portfolios and the company where investments done show a high social risk score in relation to environment and governance risk.

The regressions control the "Net Assets", similarly acknowledged as AUM, which is Morningstar equity style category. All the variables are released by MS in April 2023. For CS OLS estimation to deliver optimum results, "White (1980)" is adapted to estimate procedure to come up with the heteroscedasticity robust standard errors.

Table 1 Descriptive statistics

Variables	N	Mean	Median	Std. deviation
Sus score	64	29.68	28.58	3.08
Env score	64	4.264	4.205	1.58
Soc score	64	7.593	7.63	1.38
Gov score	64	7.007	6.82	0.88
SD three year	64	11.795	15.62	9.5
Beta three year	64	1.743	1.05	2.83
Net assets	64	6063.64	4000	8331.59

7 Findings & Discussion

This study investigates the influence of sustainability on the return volatility of MF. Utilizing a sample of 64 MF rated by Morningstar, we employ CS regression analysis and identify that mutual funds invested in portfolios with elevated governance risk are more prone to encountering heightened volatility in stock returns. Furthermore, companies demonstrating proactive measures to safeguard frontline workers, ensure customer safety, enhance healthcare benefits for employees, and forgo executive bonuses to prevent layoffs tend to experience benefits such as a more engaged and productive workforce, along with a loyal customer base during the "post-COVID-19" recovery. The trend of justifiable investing suggests a shift back to the fundamentals of stakeholder theory, asserting that commerce can engender incomes while concurrently safeguarding the earth, creating occasions for a broader population, and contributing to more unbiased and sustainable societies. These outcomes are valuable as a reference point for investors, businesses, and broader stakeholders. Business managers aiming to build resilient companies may also find this study informative (Figs. 1, 2).

Taking action to enhance "ESG" performance is crucial for companies. Poor ESG risk management can result in significant declines and underperformance in stock prices, senior management turnover, damage to reputation, negative media exposure, and increased legal litigation—all of which can be detrimental to "post-Covid-19" recovery efforts. It's imperative to annotate that this study has confines, including a comparatively trifling sample size for CSA. Additionally, "ESG scores" may exhibit bias toward larger firms that publish more data, potentially receiving better scores. Future research should focus on making progress to improve the quantity and relevance of data, providing better inputs for investment decision-making.

Table 2 shows that sustainability score, social risk score and governance risk score are positively associated with three-year standard deviation, significant at 5%

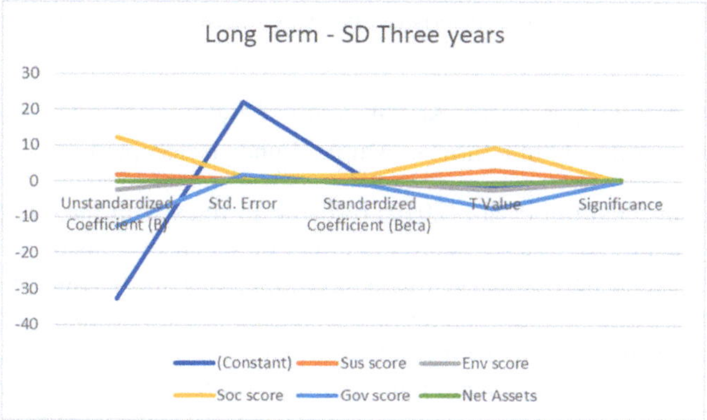

Fig. 1 II Panel A—Standard deviation—Three years

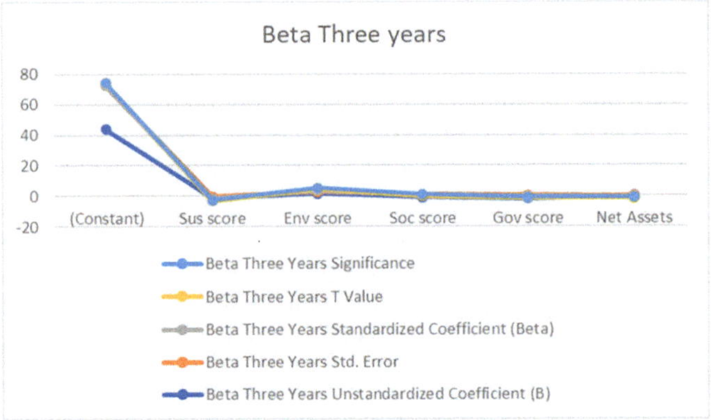

Fig. 2 II Panel beta—Three years

level. This implies that Mutual fund companies invested in portfolios with high sustainability score, social risk score and governance risk score are more likely to experience high volatility in stock returns in long term. This states that the companies having high sustainability score, social risk score and governance risk score will yield higher returns in long run. The result aligned the studies of "Orlitzky and Benjamin's (2001), Sharfman and Fernando (2008), Salama et al. (2011), Oikonomou et al. (2012), Luo and Bhattacharya (2009)" whose views states that good "ESG performance will reduce the volatility". A stakeholder-focused corporate is better in erecting its resilience in facing crises; the company may adore a unwavering sale, steady workforce turnover and worker loyalty, and good public perception, all lead to a certain level of stability in corporate. "ESG risks" can be financially material and in today's volatile environment "ESG" issues can lead to fundamental shifts in a company's management, culture and financial well-being. Those companies that are demonstrably prepared for ESG issues can better mitigate the downside risk. Hence.

H1: The sustainability score is positively associated with volatility in returns of Mutual Funds is accepted.

H2: The Environmental score, Social score, Governance score are positively associated with volatility in returns of Mutual Funds is accepted.

Pablo Duran Santomil et al., (2019) research in lines with the current result of the study, by using 2016–2018 score the effect of social responsible investment in Europe equity fund performance, was analyzed to find out whether Sustainability score showed a positive impact on performance, and it was constant with the mutual fund investments in company with good score. The findings showed that Sustainable score was evident in the flow and high-rated fund received a higher volume of fund as returns.

Table 2 Sustainability and mutual funds volatility

II Panel A—Long term—SD three years					
Model	Unstandardized coefficient		Standardized coefficient	T value	Significance
Long term - SD three year	B	Std. Error	Beta		
(Constant)	–32.724	22.004		–1.487	0.211
Sus score	1.687	0.54	0.548	3.127	0.035
Env score	–2.359	1.02	–0.393	–2.314	0.082
Soc score	12.181	1.297	1.774	9.393	0.001
Gov score	–12.496	1.647	–1.169	–7.588	0.002
Net assets	–7.136E-005	0.000	–0.063	–0.597	0.582
Beta three year	B	Std. Error	Beta		
(Constant)	43.959	28.544		1.540	0.198
Sus score	–0.997	0.700	-1.087	–1.424	0.228
Env score	1.551	1.323	0.867	1.173	0.306
Soc score	–0.720	1.682	–0.352	–0.428	0.691
Gov score	–1.843	2.136	–0.578	–0.863	0.437
Net assets	0.000	0.000	–0.419	–0.920	0.410

8 Conclusion, Limitations and Future Research

Investments are usually made on the basis of risk appetite and taking into consideration the previous performance or it can be said that the investors anticipate positivity correlational aspect among the fund past return and flow of funds. Anyway ESG investor makes a significant impact when compared with the other non-finance factor such as the need for the investments, the sizable portion which they are allocating for their investments, aging factor, return volatility, and the fee charged instead of relying on merely the past return of the fund. So it is anticipated that ESG investor may have to perform on a weaker base when compared with non- ESG fund investor. This paper examined the impact of performance of mutual funds towards achieving sustainable development goals. Making use of a sampling comprising 64Mutual Funds rating given from Morningstar in April 2023, CSRA is done, outcome states, Mutual Funds invested in portfolios with social risks, tend to show higher volatile index stock return. While companies come up with additional step to safeguard worker in the front end, safeguards customer, enhance healthcare beneficials for employee, and sacrifice executive bonus which can aid for avoidance of layoff, there by getting advantage of having engaging and production oriented workforce and loyalty in customers. Results will help for investor, business people, and wider stakeholder all over the world.

Technology company when compared with energy based sectors stock may display higher social risk. This is in comparison to other sector when this period was a period of recovery from coronavirus and so they could have experience sharp plunge in stock price, and here the technology sector stock comparatively high social risk has performed well. Before the spread of coronavirus investor focussed on the 'environment' element of ESG score. Maximum investor might have the taboo that sustainable investments were more into the changes caused for the climatic version in the planet. As depicted in Table 2 sustainability score, governance risk and social risk show impact in the mutual fund performance immediately the period of recovery period of Covid-19. This implied that ESG scores have impact on mutual fund performance immediately during the period of recovery of Covid-19. This depicts those investments in portfolio with higher social risk, governance risk and sustainability score can experience with a higher volatility in stock return. Business which is stakeholder' scentric will be resilient to face the crisis and the company could have a stability in sales, labour turnover and so on which could pave to a higher certainty in business.

The study faced certain limitations where sampling numbers taken are comparatively smaller for the cross-sectional analysis. Sustainability score has a chance of getting biased as bigger firm when they publish enormous data, they could get high score. It is found that there is a research gap in which there is scope to examine the reason a firm could chose a decision among various types of investment and also the ways a firm chose for making the various types of disclosure for those investment. It is found that the major portion of ESG data for all industry is irrelevant for monetary performance-based criteria and the relevant information it differs within various industries. So it can be concluded that there should be clarity with regard to the material information which usually vary among the countries, various sector of industry and the strategy the firm take and so on and not only that it should be given vital significance as well.

8.1 Future Research

The performance of mutual funds that focus on certain industries, like technology or healthcare, should be examined in greater detail to assess their potential for growth or resilience to changes in the economy. The analysis of fund family performance is another possible research topic. A family-level analysis to evaluate the effect of mutual fund families on performance and fund flow, similar to that conducted by (Gaspar et al., 2006), could be another expansion of this work. Future studies could examine whether funds in the same family exhibit distinct performance patterns or have benefits that set them apart from other funds. Future research could examine the effects of family-level tactics like cross-subsidization, which involves moving assets between funds in the same family to momentarily improve performance.

8.2 Managerial Implications

With regard to the contribution in reality, the inferences for business to manage ESG factors and the related risks are various. The influence of an extensive ESG gauging on volatility risk may swift managers to delineate a transparent and structured strategy. The outcome specified in Sect. 4, stated that greater systematic risk in a market of volatility may make a company to perform lower and the greater beta risk cause an upsurge in the negative market trend. Hence managers who deal with risk have to be capable to deal with the component in a more structured way through a larger communicative strategy among investors fraternity in the firm. Being more transparent makes in reality to diminish the "noise" influence which is related to scepticism. This will aid companies to diminish the perception of uncertainty with regard to the ESG performance, to meet the investors expectations. In consequent to it, the full fledged ESG rating, and the measurement of environmental scores may cause an upswing in the cost of capital of companies, as the beta coefficient is a risk factor to estimate the cost of capital, as in the capital asset pricing model (Fama & French, 2004).

Following best practices can greatly improve a company's ESG performance in addition to implementing acknowledged sustainability standards, releasing ESG reports on a regular basis, and making sure that auditing and assurance are done correctly. These procedures support continual improvement, guarantee adherence to international standards, and simplify work. Instead of considering ESG initiatives as distinct, stand-alone operations, businesses should integrate them into their main business plan to optimize their impact. This involves matching company objectives with ESG goals, which facilitates better resource allocation and decision-making. Working together with stakeholders, such as staff members, investors, clients, suppliers, and local communities, is essential to develop a successful ESG strategy. Setting clear, quantifiable, and time-bound ESG objectives is crucial for monitoring advancement and guaranteeing responsibility. Businesses can more effectively track ESG parameters and expedite reporting procedures with the help of developments in data analytics and artificial intelligence. Participating in international programs such as the "Carbon Disclosure Project" (CDP) or UN Global Compact shows a company's dedication to sustainability. Joining these networks gives access to fresh tactics and best practices in addition to increasing credibility.

8.3 Social Implications

The results of this study can help investors, fund managers, and market players make better investment decisions and add more sustainable funds to their portfolios. The incorporation of more sustainable investments into the financial products provided by institutional investors should be encouraged or even mandated by regulators and legislators. The availability of data pertaining to mutual fund ESG scores is the study's primary constraint.

References

Ali, S., Badshah, I., Demirer, R., Hegde, P., Rognone, L. (2024). Climate risk, ESG ratings, and the flow-performance relationship in mutual funds. *Global Finance Journal, 63*, 101041, ISSN 1044–0283, https://doi.org/10.1016/j.gfj.2024.101041.

Benson, K. L., & Humphrey, J. (2008). Socially responsible investment funds: Investor reaction to current and past returns. *J. Bank. Financ., 32*, 1850–1859.

Bollen, N. P. B. (2007). Mutual fund attributes and investor behavior. *J. Financ. Quant. Anal., 42*, 683–708.

Bowen, H. R. (1953). *Social responsibilities of the businessmen.* Harper & Brothers.

Brammer, S., & Pavelin, S. (2006). Voluntary environmental disclosures by large UK companies. *Journal of Business Finance and Accounting, 33*(7&8), 1168–1188.

Clarkson, M. E. (1995). A stakeholder framework for analyzing and evaluating corporate social performance. *Academy of Management Review, 20*(1), 92–117.

Dolvin, S. D., Fulkerson, J. A., & Krukover, A. (2017). *Do 'good guys' finish last? The relationship between Morningstar sustainability ratings and mutual fund performance.* Available online: https://ssrn.com/abstract=3019403. Accessed on 22 September 2017.

Dyllick, T., & Hockerts, K. (2002). Beyond the business case for corporate sustainability. *Business Strategy and the Environment, 11*, 130–141.

Friedman, F. (1970). A Friednzan Doctrine. *The New York Times.*

Gaspar, J., Massa, M., & Matos, P. (2006). Favoritism in mutual fund families? Evidence on strategic cross-fund subsidization. *The Journal of Finance, 61*(1), 73–104.

El Ghoul, S., & Karoui, A. (2017). Does corporate social responsibility affect mutual fund performance and flows? *J. Bank. Financ., 77*, 53–63.

Hartzmark, S. M., & Sussman, A. B. (2019). Do Investors value sustainability? A natural experiment examining ranking and fund flows. *The Journal of Finance., 74*(6), 2789–2837.

https://simplysustainable.com/insights/why-esg-ratings-matter-and-how-companies-use-them https://alp.consulting/esg-funds-in-india/

https://www.bajajfinserv.in/investments/mutual-fund-industry#:~:text=As%20of%20June%202024%2C%20the,61%2C15%2C582%20crore

Rajeev Kumar (2021, November) *Tremendous growth potential for ESG investing in India* https://www.financialexpress.com/money/tremendous-growth-potential-for-esg-investing-in-india-says-cfa-institutes-vidhu-shekhar-interview/2376584/

Lennsen, J. J. et al., Sustainability, Risk Management and governance: Towards an integrative approach. *Corporate Governance International Journal of Business in Society, 14*(5), 670–684 https://doi.org/10.1108/CG-07-2014-0077.

Momparler, A., Carmona, P., & Climent, F. (2024). Catalyzing sustainable investment: Revealing ESG power in predicting fund performance with machine learning. *Computational Economics.* https://doi.org/10.1007/s10614-024-10618-0

Haitham Nobanee (2020) Sustainability and risk – A Mini Review https://doi.org/10.2139/ssrn.3539058

Orlitzky, M., & Benjamin, J. D. (2001). Corporate social performance and firm risk: A meta-analytic review. *Business and Society, 40*(4), 369–396.

Santomil, P. D., et al. (2019). Does sustainability score impact mutual fund performance? *Sustainability.* https://doi.org/10.3990/su11102972

Sharfman, M. P., & Fernando, C. S. (2008). Environmental risk management and the cost of capital. *Strategic Management Journal, 29*(6), 569–592.

Pham, D. C., Do, T. N. A., Doan, T. N., Nguyen, T. X. H., & Pham, T. K. Y. (2021) The impact of sustainability practices on financial performance: empirical evidence from Sweden. *Cogent Business & Management, 8*(1). https://doi.org/10.1080/23311975.2021.1912526

Topping, N. (2012). How does sustainability disclosure drive behavior change? *Journal of Applied Corporate Finance, 24*(2), 45–48.

Vilas, P., Andreu, L., Serrano-Cinca, C. (2024). The limited role of sustainability in mutual fund investor decisions: A machine learning approach. *Expert Systems with Applications, 47*, 123288, ISSN 0957–4174, https://doi.org/10.1016/j.eswa.2024.123288

[illegible]

Impact of Hindenburg Report on the Indian Stock Market with Reference to the Adani Group of Companies

Dhanraj Sharma, Ruchita Verma, and Nihal

Abstract This research paper investigates the impact of the Hindenburg Research report published in January 2023 on the stock prices and financial performance of the Adani Group of companies. An event study methodology is employed to analyze the effect of the Hindenburg report on Adani Group stocks. Additionally, a panel data regression model is used to assess financial performance. The model included abnormal returns as the dependent variable and Size, Leverage, Liquidity, Profitability, and Tangibility as independent variables. The analysis reveals significant negative abnormal returns for Adani Group stocks following the report's release, indicating a substantial decline in stock prices. This suggests that the market reacted strongly to the allegations of fraud and manipulation. The financial performance analysis further highlights a marked adverse impact on the companies' quarterly results post-report. This research addresses a critical gap in existing literature by focusing not only on the immediate aftermath but also on the sustained effects of the Hindenburg report on Adani Group's stock performance and financial health. Investors, regulators, and policymakers can gain valuable insights from this study on the importance of monitoring and responding to corporate allegations. It emphasizes the necessity for due diligence and the potential market repercussions of financial misconduct.

Keywords Adani group · Indian stock market · Hindenburg report · Event study · Regression

D. Sharma · R. Verma (✉) · Nihal
Department of Financial Administration , Central University of Punjab, Bathinda, India
e-mail: ruchitaverma@cup.edu.in

B. Basu et al. (eds.), *Data Analytics for Decision Making towards Business Excellence*, Uncertainty and Operations Research, https://doi.org/10.1007/978-981-95-2429-7_9

1 Introduction

Corporate governance in India has long been a topic of significant discussion, particularly in light of high-profile corporate scandals that have exposed weaknesses in regulatory oversight and enforcement mechanisms. The Indian stock market experienced significant turmoil following the release of a scathing report by Hindenburg Research, a short-selling firm. The report accused the Adani Group, a prominent Indian conglomerate, headquartered in Ahmedabad (India) of engaging in fiscal misrepresentation and share manipulation (Hindenburg, 2023). Across time, Adani Group has evolved into one of the largest and most diversified business entities in India. Originating as a commodity trading enterprise, the organization has broadened its activities across various sectors, including Transportation and Logistics, Energy and Utilities, Airports, Materials etc. and has witnessed remarkable growth in recent times. Given its considerable market presence and influence, any significant event related to the Adani Group is likely to reverberate across the stock market in India. Previous studies have found that the stock market reacts significantly to the events like announcements of accounting fraud, tax evasion, stock manipulation, violation of government contracts, Mergers and Acquisitions, election, media reports etc. (Bashir et al., 2024; Song & Han, 2017; Davidson et al., 1994; Pritchard & Ferris, 2001). Some researchers also suggest that the market reactions can be weaker due to factors like weak investor protection, less efficient legal systems, institutional factors like state ownership or limited press freedom.

The financial impacts of corporate scandals are significant, with Karpoff et al. (2008) highlighting how markets react to corporate wrongdoing, resulting in substantial costs and reputational damage for firms. Ghosh (2006) examines the relationship between corporate governance and performance in India, providing context for the Adani Group's governance issues. MacKinlay (1997) outlines event study methodologies, crucial for analyzing the Adani Group's stock price reactions to the Hindenburg report. Wang et al. (2010) explore conditions conducive to corporate fraud, particularly during IPOs, which is relevant to the Adani case. Black et al. (2015) investigate how governance affects firm value in Korea, offering comparative insights for India. Balasubramanian et al. (2010) study the link between governance practices and market value in India, shedding light on the Adani Group's governance structure. Jain (2022) examined how regulatory initiatives against corporate wrongdoing in India affected the stock market. The study's findings reveal that regulatory actions serve as an effective deterrent against corporate misconduct, leading to a significantly negative impact on a firm's stock price. Dsouza's (2016) research using event studies yielded contrasting results. They found the Indian market didn't fully absorb information from quarterly earnings announcements, potentially allowing investors to profit based on new information (weak semi-strong form inefficiency). This research paper focuses on the event of release of the Hindenburg Report, which made serious allegations about the Adani Group's financial practices and corporate governance.

Understanding how the Indian stock market reacted to this information and its impact on the financial performance of Adani Group companies is crucial for evaluating market efficiency and potential corrective measures. Insights from behavioral finance are integrated to explain the immediate, possibly overreactive, stock market response following the Hindenburg report. Supplementing the study with investor sentiment data, drawn from social media or news analytics, offers a richer perspective on market reactions and the role of public perception in driving stock price changes. By building upon this foundation, we propose the below hypotheses:

H1: Hindenburg Research report has no significant impact on the stock prices of the Adani Group of companies.

H2: Hindenburg Research report has no significant impact on the financial performance of the Adani Group of companies.

1.1 Hindenburg Report on Adani Group

The Hindenburg report released on January 24, 2023 accused the Adani Group of fiscal misrepresentation and share manipulation spanning several years. Hindenburg presented evidence suggesting deceptive accounting practices and artificial inflation of stock prices. The report further claimed the Adani Group implemented strategies to fabricate or illegitimately inflate revenue while diverting funds from publicly listed companies. Hindenburg's accusations extended to the Adani family's use of offshore shell entities in tax havens for potentially illicit activities like corruption, money laundering, and misappropriation of taxpayer funds. The report concluded that the Adani Group's valuation could be inflated by as much as 85%, citing concerns about excessively high valuations based on several traditional metrics (Hindenburg, 2023). Major allegations raised against Adani Group in Hindenburg report were:

- Inflated Stock Value: The report claimed Adani shares were significantly overvalued based on traditional financial ratios.
- Reliance on Debt: Five out of seven key listed companies reportedly had a current ratio below 1, indicating a potential overreliance on debt.
- Promoters' Stock Pledging: The company's founders (promoters) reportedly borrowed against their shares, suggesting additional debt burdens.
- Excessive Promoter Influence: Allegations suggested a significant portion of publicly available shares were managed by front companies associated with the Adani Group, potentially inflating promoter influence beyond their direct ownership stake.

The Adani Group vehemently denied the allegations in a 413-page response, characterizing the Hindenburg report as a "malicious" and "baseless" attempt to create a false market through misinformation (Adani Group, 2023). They emphasized their commitment to India's economic progress and threatened legal action to defend their reputation.

1.2 Stock Price Performance After Hindenburg Report

The Figs. 1, 2, 3, 4, 5, 6, 7, 8, 9 and 10 illustrate the cumulative returns of Adani Group companies since the publication of the Hindenburg report. The analysis is divided into three timeframes: 0–33 days, 34–66 days, and 67–100 days post-publication. In the initial period post-publication (0–33 days), the data reveals a significant negative trend in the cumulative returns across all Adani Group companies. This sharp decline suggests an immediate adverse market response to the allegations of financial impropriety and stock manipulation outlined in the Hindenburg report. This phase is indicative of heightened investor anxiety and a rapid re-evaluation of the companies' valuations.

The second timeframe (34–66 days) highlights varying recovery trajectories and continued decline among the different companies. Certain firms, such as Adani Enterprises and Adani Ports & SEZ, demonstrate partial recovery, suggesting a potential reassessment of the initial shock by market participants based on new information or corporate responses. However, other companies, such as Adani Green Energy

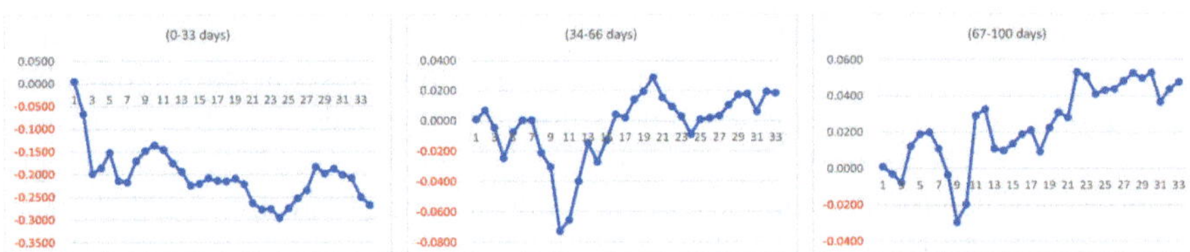

Fig. 1 ACC

Fig. 2 Ambuja cement

Fig. 3 Adani enterprises

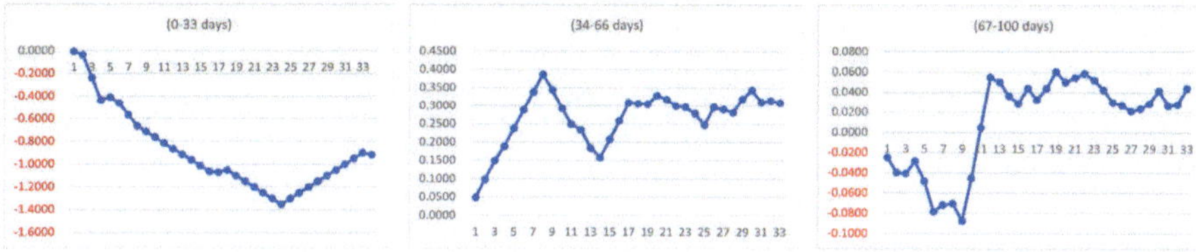

Fig. 4 Adani green energy

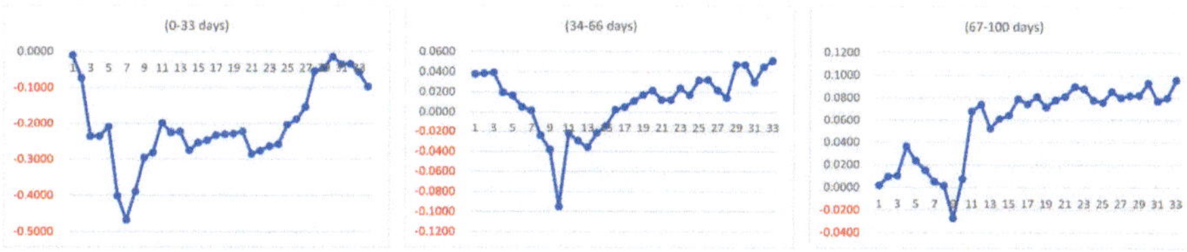

Fig. 5 Adani ports & sez

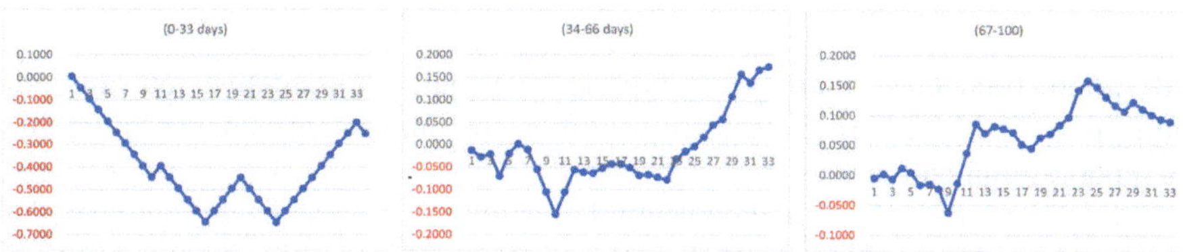

Fig. 6 Adani power

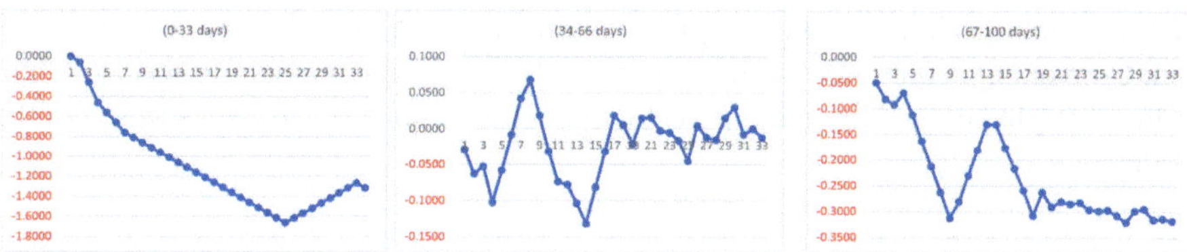

Fig. 7 Adani total gas

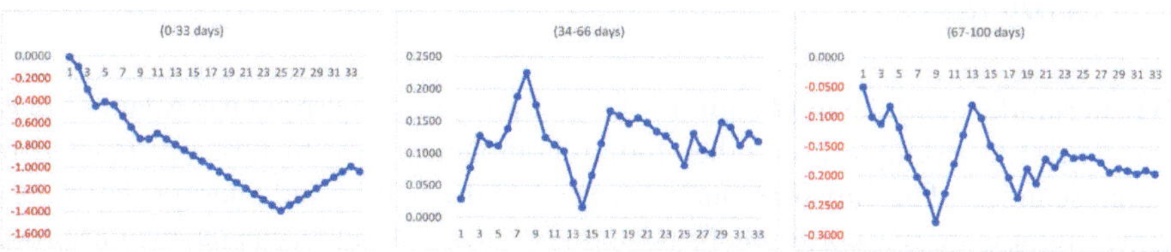

Fig. 8 Adani energy solutions

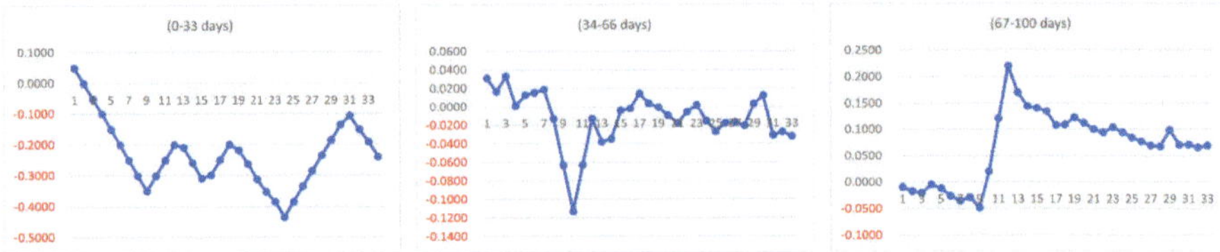

Fig. 9 Adani Wilmar

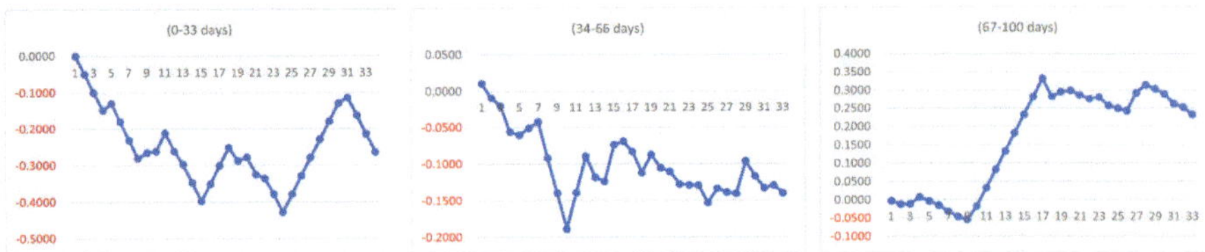

Fig. 10 NDTV (*Source* Authors)

and Adani Total Gas, continue to experience negative cumulative returns, indicating potentially sustained negative sentiment or unresolved concerns regarding the report's implications.

The final period's (67–100 days) cumulative returns provide insight into whether the companies have achieved stabilization, continued to decline, or exhibited signs of recovery. This phase is crucial for understanding the long-term ramifications of the Hindenburg report. For example, stabilization or recovery for companies like ACC or Ambuja Cement could imply an overestimation of the initial negative reaction. Conversely, continued declines, potentially observed in Adani Power or Adani Wilmar, might suggest lingering investor concerns or unresolved issues raised in the report.

1.3 Historical Reactions of Indian Companies to Corporate Scandals

To better understand the implications of the Hindenburg report on the Adani Group, it is crucial to examine how other Indian companies have historically reacted to similar scandals. The Satyam scandal, as examined by Turka and Sharma (2013), is a prime example of how corporate fraud and governance failures can have widespread repercussions, prompting regulatory bodies like SEBI to tighten regulations and enforcement. According to Varma (1997), the governance framework in India has historically been marred by issues of transparency and accountability, which have often led to conflicts of interest and fraudulent practices. The Satyam scandal had immediate and long-term financial repercussions on the company. The immediate market reaction was a sharp decline in Satyam's stock price, similar to the Adani Group's experience. However, over the long term, Satyam faced significant challenges in regaining investor confidence, ultimately leading to its acquisition by Tech Mahindra. Previous research by Deb and Mukherjee (2008) highlights how market manipulation can distort market efficiency, leading to significant financial losses for investors and undermining confidence in the market.

Another example is the ICICI Bank scandal in 2018, where allegations of conflict of interest and corruption led to an immediate drop in stock prices. However, the long-term impact included regulatory scrutiny, changes in corporate governance, and efforts to restore public trust. These cases illustrate that while immediate financial impacts are often severe, the long-term effects depend on the company's response, regulatory actions, and changes in investor sentiment. The regulatory frameworks in India, primarily led by SEBI, are crucial in addressing corporate allegations through enhanced transparency and governance reforms. Despite these efforts, enforcement challenges remain, especially with powerful conglomerates like the Adani Group. Investor sentiment significantly impacts stock volatility, with negative news often causing sharp declines and increased trading volumes. The Hindenburg report on Adani Group exemplifies this, highlighting the critical role of regulatory responses in stabilizing markets. Effective regulatory scrutiny is essential to restore investor confidence and mitigate market volatility.

1.4 Global Comparisons

The impact of the Hindenburg report on the Adani Group can be contextualized by comparing it with similar global cases of corporate misconduct. The Enron scandal in 2001, where Enron used complex accounting practices to hide debt and inflate profits, led to a drastic fall in stock prices and loss of investor confidence, similar to the immediate reaction to the Adani case. This resulted in the introduction of the Sarbanes–Oxley Act to enhance corporate transparency and accountability. The Tesla-Musk controversies, particularly Elon Musk's 2018 tweet about taking Tesla private, led to SEC charges for fraud and significant volatility in Tesla's stock prices, underscoring the importance of regulatory oversight and transparency (Gow et al., 2019). In 2020, the Wirecard scandal in Germany, where €1.9 billion were found missing from its accounts, led to a collapse in its stock price and the company's insolvency, highlighting significant regulatory lapses and prompting calls for reforming financial oversight mechanisms. These global cases illustrate how corporate misconduct can have immediate and long-term financial impacts, emphasizing the need for robust regulatory frameworks and transparency to protect investors and maintain market integrity.

1.5 Statement of the Problem

The release of the Hindenburg Research report has raised significant concerns regarding the transparency and integrity of corporate practices within major conglomerates. The report's allegations led to immediate market reactions, not only affecting Adani Group's stock prices but also causing ripples across the Indian stock market. The problem lies in understanding the extent to which such investigative reports can impact market efficiency, investor behavior, and the financial stability of major corporations. Specifically, this study seeks to determine the immediate and long-term effects of the Hindenburg report on the Adani Group's stock performance and financial health. Additionally, it aims to evaluate whether the Indian stock market efficiently incorporated the information presented in the report, as posited by the Efficient Market Hypothesis (EMH). Addressing this problem is crucial for several reasons. First, it provides insights into how significant negative information about a major conglomerate is processed by the market. Second, it highlights the need for robust regulatory frameworks to prevent market manipulation and ensure corporate accountability. Finally, understanding the impact of such events can help investors and policymakers develop strategies to mitigate risks associated with corporate misconduct and maintain market integrity.

2 Methodology

2.1 Sample and Data Source

Only secondary data were used for this study. Stock price data was examined for two sets of stocks: the companies in Table 1 and the Nifty 50 Index. This analysis included daily closing prices, with a specific focus on the date the Hindenburg Research report was announced. All data were retrieved from Yahoo finance and NSE websites. This study also adopted an 8-quarters financial data approach (four quarters pre- and four quarters post-release of the Hindenburg report) to shed light on the financial repercussions the Hindenburg report had on the Adani Group. Quarterly results were retrieved from company websites (Table 2).

2.2 Event Study Methodology

Financial researchers often use event studies to measure the effect of an event (like a news announcement) on a company's stock price (Mallesha & Archana, 2023). This methodology aligns with the Efficient Market Hypothesis (Fama, 1965, 1970), which posits that investors integrate new information into stock prices, affecting both

Table 1 Major Timeline of the case

Date	Timeline
Jan 24, 2023	Hindenburg Research releases a report accusing the Adani Group of accounting fraud, stock manipulation, and overleveraging
Jan 29, 2023	Adani Group issues a 413-page rebuttal to the Hindenburg report, denying all allegations. The rebuttal claims the report is "malicious" and "baseless."
Feb 01, 2023	Adani Enterprises cancels its fully subscribed FPO opting to refund proceeds in collaboration with FPO banks
Feb 2, 2023	Adani Group's market cap drops below $100 billion as shares decline, while Credit Suisse and Citigroup's wealth arm halt loans to the conglomerate
Feb 6, 2023	Fitch Ratings downgrades Adani Green Energy's rating due to concerns about the Hindenburg report
Feb 27, 2023	Adani Group's fixed-income roadshow in Singapore to boost investor confidence
Feb 28, 2023	Adani Group's fixed-income roadshow in Hong Kong to boost investor confidence
Mar 01, 2023	Adani Group secures an $800 million debt facility to refinance Adani Green Energy's $750 million bond due in September 2024. The conglomerate also announces securing $3 billion in credit from a sovereign wealth fund
May 17, 2023	Adani Group loses a $535 million arbitration against NHAI
Jun 14, 2023	Adani Green Energy's shares plunged after Moody's Investors Service downgraded its rating due to increased debt and project delays
Aug 25, 2023	SEBI's preliminary report finds no evidence of stock manipulation by Adani Group but raises concerns about certain accounting practices
Oct 26, 2023	Rosenkranz PLLC, a US law firm, files a class-action lawsuit against Adani Group for investors' losses post-Hindenburg report
Nov 28, 2023	Supreme court said the Hindenburg Research report should not be treated as a statement of truth
Dec 05, 2023	Bloomberg reported US agency concluded accusations against the Adani Group of Companies weren't relevant
Jan 3, 2024	SC denies transferring the probe to special investigation team, instructing SEBI to conclude its investigation into two pending cases within 3 months

Sources The Economic Times (Feb 01, 2023), The New Indian Express (26 Aug 2023) The Hindu Business Line (January 03, 2024), Business Insider (Jun 06, 2023), Reuters (August 28, 2023)

Table 2 Major listed Adani Group of companies, their industry and year of establishment

Code	Adani Group Companies	Industry	Year of Establishment
C1	ACC	Cement	1936, acquired in 2022
C2	Ambuja Cement	Cement	1966, acquired in 2022
C3	Adani Enterprises	Trading	1988
C4	Adani Green Energy	Renewable Energy	2015
C5	Adani Ports & Sez	Transport Infrastructure	2006
C6	Adani Power	Power Generation	2009
C7	Adani Total Gas	Gas Distribution	2016
C8	Adani Energy Solutions (Adani Transmission)	Power Generation and Distribution	2012
C9	Adani Wilmar	Consumer Foods	1999
C10	NDTV	TV Broadcasting	1988, acquired in 2022

individual stocks and the broader market. The event study approach involves calculating expected and actual returns within a designated event window. By analyzing the difference (abnormal return), we can isolate how much the event specifically impacted stock performance, independent of general market fluctuations. In addition to the traditional event study methodology, behavioral finance insights are incorporated to understand the immediate, possibly overreactive, market response to the Hindenburg report. Investor sentiment data, drawn from social media and news analytics, is also analyzed to provide a comprehensive view of market reactions.

The event study methodology involves the following steps:

1. Event Definition: Identifying the event, which in this case is the release of the Hindenburg Research report on January 24, 2023.
2. Estimation Window: Selecting a period before the event to estimate the normal performance of the stock. Typically, an estimation window of 100 days is used.
3. Event Window: Defining the period around the event to study the impact. This study uses a 100-day window around the event date.
4. Modelling Expected Returns: Using a market model to predict the expected returns during the event window. The market model assumes a linear relationship between the return of the stock and the return of the market.
5. Calculating Abnormal Returns: The abnormal return (AR) is the difference between the actual return and the expected return.
6. Statistical Significance Testing: Conducting t-tests to determine whether the abnormal returns are statistically significant.

The mathematical formulation of the market model is as follows:

$$E(R_{it}) = \alpha_i + \beta_i R_{mt} + \varepsilon_{it} \text{ for } i = 1 \ldots N \tag{1}$$

To determine expected returns $\alpha 1$ and $\beta 1$ values are needed. This is achieved through a basic regression model, estimating security returns by incorporating actual market returns, $R_{mt.}$

$$\text{Expected Returns} = E(R_{it}) = \alpha_i + \beta_i R_{mt} \tag{2}$$

The calculation of ARs is conducted using the following approach:

$$AR_{it} = e_t = R_{it} - E(R_t) \tag{3}$$

The methodology utilized to compute AARs is outlined as follows:

$$AAR_{it} = \frac{\sum_{i-1}^{N} AR_{it}}{-N} \tag{4}$$

Over time, ARs are computed to derive the CARs in the following manner:

$$CAR[t_1, t_2] = \sum_{t_1}^{t_2} AR \tag{5}$$

Then, Statistical significance of abnormal returns is also determined. A significant t-statistic suggests the event impacts returns, with the direction indicating positivity or negativity.

$$t = \frac{AAR}{\sigma(AAR)} \tag{6}$$

t statistics for CAAR are computed using the formula:

$$t = \frac{CAAR}{\sigma(CAAR)} \tag{7}$$

2.3 *Regression Model*

The regression methodology employed in this research utilized a panel data regression model to analyze the relationship between financial variables and abnormal returns. The analysis utilized a panel data set comprising observations of 8-quarter years and 10 cross-sections, encompassing four quarters pre- and four quarters post-release of the Hindenburg report with a total of 4884 observations. The model includes Size, Leverage, Liquidity, Profitability, and Tangibility as independent variables. Statistical tests, including normality, stationarity and heteroskedasticity tests, will ensure the analysis' validity and reliability. Additionally, robust standard error estimation will be employed to provide a nuanced picture of how firm characteristics influences abnormal returns and to enhance the accuracy of parameter estimates and statistical inferences (Table 3).

Table 3 Variable description

Code	Variable	Description	Source
AR	Abnormal Return	Log value of AR, derived by subtracting the expected return from the actual return	(Mallesha & Archana, 2023; Frendy & HU Dan, 2014)
Size	Total Asset	Log value of total assets	(Zhu et al., 2017; Bashir et al., 2024; Rahman et al, 2021)
LEV	Leverage	Ratio of total debt to total assets	(Zhu et al., 2017)
LIQ	Liquidity	Ratio of cash and short-term investments to total assets	(Kartal Demirgüneş, 2016; Krüger, 2015)
PROF	Profitability	Ratio of net income to total assets	(Zhu et al., 2017; Bashir et al., 2024)
TAN	Tangibility	Ratio of fixed assets to total assets	(Yüksel İltaş & Kartal Demirgüneş, 2020; Irungu et al., 2018)

Following Model is constructed to measure the relationship of dependent and independent variables:

$$AR_i = Y_0 + Y_1 SIZE + Y_2 LEVERAGE + Y_3 LIQUIDITY + Y_4 PROFITABILITY + Y_5 TANGIBILITY$$

where AR_i is abnormal returns over the window period. The independent variables consist of idiosyncratic firm characteristics selected from existing literature (Kolaric & Schiereck, 2016; Krüger, 2015; Rahman et al., 2021)

3 Results

In the results section, a comprehensive analysis and interpretation is presented through a series of tables and tests that support the findings of the study.

3.1 Results Based on Event Study

3.1.1 Immediate Financial Impacts

The immediate financial impacts of the Hindenburg report on the Adani Group were significant. On January 24, 2023, the day the report was released, Adani Group stocks experienced a decline. The negative impact of the Hindenburg report becomes more pronounced from day + 1 onwards. The immediate market reaction was characterized by heightened volatility and a rapid decline in stock prices across all Adani Group companies. Behavioral finance insights suggest that this overreaction may have been driven by investor sentiment amplified by media coverage. Analysis of social media and news sentiment data reveals a strong correlation between negative sentiment and stock price declines (Table 4).

The pre-event period shows mostly insignificant abnormal returns, indicating no substantial anticipation of the report's impact by the market. The only exception is Adani Enterprises on day -3 with a significant AR of –0.0360**. On the event day (day 0), all companies except ACC, Adani Power, and Adani Wilmar experienced negative abnormal returns, although these were not statistically significant. In the post-event window, the negative impact of the Hindenburg report becomes more pronounced. Day + 1 and + 2 show significant negative ARs for multiple companies. For example, ACC (–0.0615*** and –0.1173***), Ambuja Cement (–0.0678*** and –0.1603***), and Adani Ports & SEZ (–0.0462*** and –0.1411***) all show substantial declines. The pattern continues on day + 3, where Adani Enterprises reports a significant positive AR (0.0408**), but Adani Total Gas (–0.2064***) and Adani Transmission (–0.1566***) continue to experience significant losses. This trend continued on day 4, with ACC, Ambuja Cement, and Adani Total Gas experiencing significant abnormal returns (negative for Adani Total Gas), while others remained insignificant. Finally, on day 5, all companies exhibited negative abnormal returns, with significance for ACC, Ambuja cement, Adani Enterprises, Adani Green Energy, Adani Ports & SEZ, and Adani Total Gas. These comparisons underscore that firm-specific attributes and external shocks significantly influence stock market reactions, supporting the Efficient Market Hypothesis (EMH) where new information is rapidly reflected in stock prices.

Kothari and Warner (2007) and Brown and Warner (1985) observed that significant market events or news announcements often lead to shifts in ARs, reflecting the market's rapid incorporation of new information. MacKinlay (1997) noted that the market's reaction can vary significantly depending on the nature and perceived impact of the event. In the context of the Adani Group, the observed ARs on key dates and one day after the event further illustrate how specific corporate events can trigger substantial market reactions, both positive and negative, consistent with the Efficient Market Hypothesis (EMH). Table 5 presents the abnormal returns (ARs) for key dates and one day after the event, highlighting both statistically significant and insignificant market reactions. For example, on January 25, 2023, ACC, Ambuja Cement, and Adani Ports & SEZ experienced statistically significant negative ARs, indicating an adverse market reaction to the event on that date. Similarly, on February 1, 2023, significant negative ARs were observed for ACC, Ambuja Cement, Adani Enterprises, Adani Green Energy, and Adani Total Gas, underscoring the negative impact of these events.

Conversely, positive abnormal returns were also noted for certain companies on different dates. On November 28, 2023, all companies except ACC exhibited statistically significant positive ARs. Additionally, on December 5, 2023, all ten companies showed significant positive ARs, suggesting a favourable market response. These findings align with the broader literature on market reactions to corporate events (see eg: Kothari & Warner, 2007; Brown & Warner, 1985; MacKinlay, 1997).

Table 4 ARs and AARs in event window of –5, 5

	–5	–4	–3	–2	–1	0	1	2	3	4	5
C1	–0.0155	0.0016	0.0135	–0.0112	–0.019	0.0047	–0.0615 ***	–0.1173 ***	0.01	0.0324 **	–0.06 ***
C2	–0.0084	–0.0013	–0.0009	–0.0003	–0.0381	–0.0057	–0.0678 ***	–0.1603 ***	0.0149	0.0333 *	–0.1663 ***
C3	–0.0101	–0.0236	–0.036 **	0.0007	–0.016	–0.0016	–0.0015	–0.1663 ***	0.0408 **	0.0237	–0.2817 ***
C4	0.0255	–0.0402	–0.0664	0.0138	–0.0292	–0.0095	–0.0194	–0.1849 ***	–0.2044 ***	0.0268	–0.055 **
C5	–0.0157	–0.0069	–0.0094	0.0036	–0.0145	–0.0107	–0.0462 ***	–0.1411 ***	–0.0039	0.0247	–0.1887 ***
C6	0.0165	–0.0245	–0.0046	–0.0078	–0.0181	0.0006	–0.0378	–0.0334	–0.0581 *	–0.0556	–0.0511
C7	0.0178	–0.0225	0.0086	0.024	–0.014	–0.0056	–0.0473	–0.1833 ***	–0.2064 ***	–0.1041 ***	–0.1 ***
C8	–0.0036	–0.014	0.0102	0.0044	0.0094	–0.0098	–0.0766 ***	–0.1841 ***	–0.1566 ***	0.0358	–0.0275
C9	0.0066	–0.0118	–0.0164	–0.0102	–0.0165	0.0468	–0.0549	–0.0557	–0.0517	–0.0521	–0.0527
C10	–0.0121	0.0002	–0.0407	–0.027	0.003	–0.0051	–0.0457	–0.0429	–0.0568	0.0131	–0.053
AAR	0.0001	–0.0143	–0.0142	–0.0011	–0.0153	0.0004	–0.0459	–0.1269	–0.0672	–0.0022	–0.1036

Source Authors

Note This table shows the Abnormal Returns (AR) and Average Abnormal Return (AAR) for the event window (–5,5). *, **, *** denote statistical significance at the 10%, 5% and 1% level, respectively

Table 5 ARs on Key timelines related to event

	C1	C2	C3	C4	C5	C6	C7	C8	C9	C10
Jan 24 2023	–0.0615 ***	–0.0678 ***	–0.0015	–0.0194	–0.0462 ***	–0.0378	–0.0473	–0.0766	–0.0549	–0.0457
Feb 01, 2023	–0.0600 ***	–0.1663 ***	–0.2817 ***	–0.0550 *	–0.1887 ***	–0.0511	–0.1 ***	–0.0275	–0.0527	–0.053
Feb 2, 2023	–0.0031	0.0541 ***	–0.2698 ***	–0.1014 ***	–0.0661 ***	–0.0541	–0.1028 ***	–0.1015 ***	–0.0523	–0.0546
Feb 6, 2023	0.0264	0.0196	–0.0053	–0.0465	0.0999 ***	–0.0481	–0.047	–0.0963 ***	–0.0533	0.0142
Feb 27, 2023	–0.0169	–0.0421 **	–0.0904 ***	–0.0473	0.0109	–0.0489	–0.0480 *	–0.0472 *	–0.053	–0.0517
Feb 28, 2023	0.026	0.0402 **	0.1465 ***	0.0536	0.0608 ***	0.0519	–0.0469	–0.0462 *	0.0466	0.0488
Mar 01,2023	0.0132	0.0246	0.1319 ***	0.0393	0.004	0.0339	0.035	0.0387	0.0495	0.0387
May 17,2023	–0.0093	0.0172	0.0101	0.0057	0.004	–0.0049	–0.0461	–0.0228	0.0033	–0.0158
Jun 14, 2023	0.0017	0.009	–0.0084	–0.0015	–0.0014	–0.017	–0.0191	–0.0218	–0.0032	0.0155
Aug 25, 2023	–0.0086	–0.0278	–0.0119	0.0005	–0.011	–0.0078	–0.0078	–0.018	–0.009	–0.0142
Oct 26, 2023	0.0172	0.0071	–0.008	–0.0192	0.0204	0.0605 *	–0.0164	0.0415	–0.0141	–0.0062
Nov 28, 2023	0.0215	0.0351 *	0.0791	0.1157	0.0458	0.1123	0.1906	0.1793	0.0982	0.1126
			***	***	***	***	***	***	***	**
Dec 05, 2023	0.0735	0.064	0.1546	0.1897	0.1412	0.1425	0.1855	0.1891	0.0989	0.1738

(continued)

Table 5 (continued)

	C1	C2	C3	C4	C5	C6	C7	C8	C9	C10
	***	***	***	***	***	***	***	***	***	***
Jan 3, 2024	0.0116	0.0146	0.0302 *	0.0639 *	0.0238	0.0543	0.1036 ***	0.1193 ***	0.0355	0.0354

Source Authors
Note This table shows the Abnormal Returns (AR) for the key time points and one day after the event. *, **, *** denote statistical significance at the 10%, 5% and 1% level, respectively

3.1.2 Long-Term Financial Impacts

The long-term financial impacts of the Hindenburg report were also profound. Over the 100-day event window, the cumulative abnormal returns (CAR) showed a varied pattern of partial recovery and continued decline among different Adani Group companies. For example, while Adani Enterprises and Adani Ports & SEZ showed signs of recovery in the mid-term window (21–40 days), companies like NDTV continued to experience negative cumulative returns, indicating sustained negative sentiment and unresolved concerns. All companies experienced negative CAR in 41–60 event window. This suggests that the initial shock had lasting effects on investor confidence and the financial stability of the Adani Group (Table 6).

Window 0–20: All companies except Adani Wilmar and NDTV experienced a statistically significant drop in stock price (negative CAR) at the 1% level. Adani Wilmar and NDTV experienced decline at 5% level significance. This indicates a substantial negative market reaction to the Hindenburg report.

Table 6 CARs and CAARs of Adani Group of Companies

	0–20	21–40	41–60	61–80	81–100
C1	–0.2461 (0.0022) ***	0.0103 (0.8824)	–0.0491 (0.4814)	–0.0137 (0.8439)	–0.0042 (0.9514)
C2	–0.3568 (0.0002) ***	0.1096 (0.1729)	–0.0262 (0.7390)	0.0313 (0.6905)	0.0118 (0.8803)
C3	–0.7599 (0.0000) ***	0.2627 (0.0045) ***	–0.0866 (0.3014)	0.2426 (0.0077) ***	–0.1553 (0.0719) *
C4	–1.2065 (0.0000) ***	0.6163 (0.0000) ***	–0.1282 (0.2765)	–0.0099 (0.9319)	–0.0554 (0.6337)
C5	–0.2520 (0.0042) ***	0.2175 (0.0102) **	–0.0488 (0.5297)	0.0414 (0.5935)	–0.0143 (0.8530)
C6	–0.5521 (0.0003) ***	0.1754 (0.1728)	–0.0943 (0.4555)	0.0802 (0.5250)	–0.1341 (0.2924)
C7	–1.4910 (0.0000) ***	0.1551 (0.1604)	–0.1714 (0.1230)	–0.2304 (0.0429) **	–0.2839 (0.0150) **
C8	–1.1945 (0.0000) ***	0.3304 (0.0038) ***	–0.1667 (0.1126)	–0.1575 (0.1327)	–0.1622 (0.1221)
C9	–0.3660 (0.0047) **	0.0410 (0.7192)	–0.0715 (0.5322)	0.0908 (0.4289)	–0.1153 (0.3175)
C10	–0.4059 (0.0204) **	–0.0622 (0.6970)	–0.2275 (0.1644)	0.0605 (0.7049)	–0.0822 (0.6072)
CAAR	–0.6831***	0.1856**	–0.1070	0.0135	–0.0995

Source Authors
Note This table shows the Cumulative Abnormal Returns (CAR) (t-statistics within the Parenthesis) of the Adani Group of companies. The windows are made into five categories, those are 0–20, 21–40, 41–60, 61–80, and 81–100. *, **, *** denote statistical significance at the 10%, 5% and 1% level, respectively

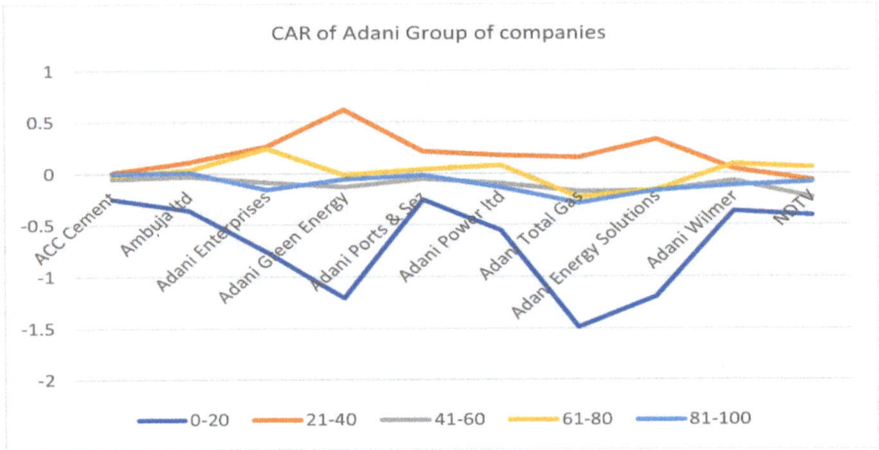

Fig. 11 CAR of Adani group of companies (*Source* Authors)

Window 21–40: There were signs of partial recovery for most companies. All except NDTV saw positive CAR, with some (Adani Enterprises, Adani Green Energy, Adani Energy Solutions, Adani Ports & SEZ) experiencing statistically significant positive returns.

Window 41–60: All companies experienced insignificant negative CAR, suggesting a period of relative stability in abnormal returns.

Window 61–80: Adani Enterprises displayed a significant positive CAR at the 10% level, while Adani Total Gas had a significant negative CAR at the 5% level.

Window 81–100: All companies except Ambuja Cement experienced negative CAR. Adani Enterprises remained significantly positive (1% level), and Adani Total Gas's negative CAR reached 5% significance. These findings suggest a complex post-event response. While an initial significant negative impact is evident, a partial recovery followed in the medium-term window (21–40 days). The later windows (post-day 40) show some positive signs for Adani Enterprises but also continued negative sentiment for Adani Total Gas and a return to negative CARs for most companies (Fig. 11).

3.2 *Results Based on Regression Model*

3.2.1 Descriptive Statistics of the Variables and Assumption Tests

Table 7 presents information on abnormal returns, company size, leverage, liquidity, profitability, and tangibility, highlighting the variability and range of these important financial metrics. The average abnormal return of approximately –0.2% suggests a

Table 7 Descriptive statistics

Variable	Obs	Mean	Std. Dev	Min	Max
AR	4884	–.002	0.031	–0.282	0.192
SIZE	4884	4.444	0.343	3.646	4.902
LEV	4884	0.236	0.212	0.001	0.675
LIQ	4884	0.057	0.126	0	0.483
PROF	4884	0.014	0.021	–0.012	0.112
TAN	4884	0.183	0.188	0	0.685

Source Authors

slight negative deviation from expected returns, with relatively low volatility indicated by a standard deviation of 0.031. The companies exhibit an average size of 4.444, with a diverse range from 3.646 to 4.902, reflecting varying company sizes. Leverage averages at 0.236, with a range from 0.001 to 0.675, indicating diverse debt-to-equity ratios. Liquidity levels average at 0.057, with a range from 0 to 0.483, showcasing varying liquidity positions. Profitability averages at 0.014, with a range from –0.012 to 0.112, indicating diverse profitability levels. Tangibility averages at 0.183, with a range from 0 to 0.685, reflecting varying levels of tangible assets (Table 8).

The normality test results reveal important information about the distribution of the standardized residuals. The skewness value of 0.331961 suggests a slight right skewness, while the kurtosis value of 12.11043 indicates heavy-tail in the data. The Jarque–Bera test statistic of 16,980.21 with a *p*-value of 0.00000 indicates a departure from normality, emphasizing the need for robust statistical methods (Table 9).

The stationarity test provides evidence regarding the presence of unit roots in the data. The statistically significant test results, with probabilities close to zero, reject the null hypothesis of unit roots, indicating stationarity in the data series (Table 10).

Table 8 Results of normality test

Mean	Median	Max	Min	Std. Dev	Skewness	Kurtosis	JB stat
1.59e–17	–0.001540	0.195049	–0.279475	0.031452	0.331961	12.11043	16,980.21 (0.00000)

Source Authors

Table 9 Results of stationarity test

Ho: All panels contain unit roots	Inverse chi-squared (20) P	Inverse normal Z	Inverse logit t (54) L*	Modified inv. chi-squared Pm
Statistic	720.8731	–25.6963	–63.4418	110.8178
Prob.**	0.0000	0.0000	0.0000	0.0000

Source Authors

Table 10 Results of Breusch-Pagan / Cook-Weisberg test for heteroscedasticity

Ho: Constant variance	chi2(1)	Prob > chi2
Variables: fitted values of AR		
	2.21	0.1374

Source Authors

The Breusch-Pagan / Cook-Weisberg test results, assesses the presence of varying levels of variance in the data. The results show that the test does not reject the null hypothesis of constant variance, suggesting homoskedasticity in the data.

3.2.2 Robust Standard Error

The robust standard error analysis reveals a nuanced picture of how firm characteristics influences abnormal returns. Pre-event, profitability emerges as the strongest driver (coefficient = 4.542, $p = 0.000$), with higher profitability linked to significantly positive abnormal returns. Conversely, leverage (coefficient = –0.157, $p = 0.006$), liquidity (coefficient = –0.699, $p = 0.000$), and tangibility (coefficient = –0.256, $p = 0.009$) are all negatively associated with abnormal returns, suggesting that companies with higher leverage, lower liquidity, and greater tangible assets tend to experience lower abnormal returns before the event. Company size (coefficient = 0.026, $p = 0.413$) appears to have no significant impact (Table 11).

Post-event, the importance of profitability diminishes slightly (coefficient = 1.272, $p = 0.073$), though it remains marginally positive. Liquidity retains its significance (coefficient = –0.663, $p = 0.000$), with lower liquidity linked to more negative abnormal returns. Leverage (coefficient = 0.012, $p = 0.856$) and tangibility (coefficient = –0.042, $p = 0.669$) lose their pre-event influence, suggesting they don't significantly impact post-event abnormal returns. In overall, Company size and Tangibility remains least significant. Similar studies have found that firm-specific factors significantly influence abnormal returns in the wake of major market events. Agarwal et al., (2021) in their study on the impact of COVID-19 on Indian hotel and tourism stocks, observed significant abnormal returns influenced by firm characteristics such as liquidity and profitability. Likewise, Basdas and Oran (2014) highlighted the sensitivity of abnormal returns to firm leverage and liquidity during financial crises in Turkey. Moreover, a study by Balyuk and Fedyk (2023) on U.S. firms' responses to geopolitical events showed that firm-specific attributes, including leverage and profitability, significantly impacted stock returns, aligning with our observations.

Table 11 Results of robust standard error model

Abnormal Returns	Pre-Event		Post Event		Overall	
	Coef	Robust std. Err	Coef	Robust std. Err	Coef	Robust std. Err
_cons	−1.984994	0.1388018	−1.290922	0.1601197	−1.673132	0.1049618
		(0.000) ***		(0.000) ***		(0.000) ***
SIZE	0.0256498	0.0312964	−0.1398807	0.035798	−0.0458383	0.0235719
		−0.413		(0.000) ***		(0.052) *
LEV	−0.157195	0.0567258	0.0117831	0.0650486	−0.1278853	0.0396811
		(0.006) ***		−0.856		(0.001) ***
LIQ	−0.6987676	0.1116252	−0.6627524	0.1127528	−0.7331796	0.0760613
		(0.000) ***		(0.000) ***		(0.000) ***
PROF	4.541817	1.096461	1.272012	0.7098316	2.351257	0.5599312
		(0.000) ***		(0.073) *		(0.000) ***
TAN	−0.2555202	0.0974156	−0.0419868	0.0980769	−0.1233949	0.0675538
		(0.009) ***		−0.669		(0.068) *
F statistics	F(5, 2578) = 27.10		F(5, 2294) = 11.99		F(5, 4878) = 33.51	
	0		0		0	
R-squared	0.0528		0.0255		0.0341	

Source Authors

4 Conclusion

This paper explored the effect of the Hindenburg report on the share prices of the Adani Group companies. We have employed an event study approach within a 100-day window, calculating abnormal returns and their significance, and panel data regression comprising observations of 8-quarter years and 10 cross-sections, encompassing four quarters pre- and four quarters post-release of the Hindenburg report with a total of 4884 observations. To ensure robustness, we employed normality, stationarity, and heteroskedasticity tests along with robust standard error estimation. The analysis reveals significant negative abnormal returns for Adani Group stocks following the report's release, indicating a substantial decline in stock prices. This suggests that the market reacted strongly to the allegations of fraud and manipulation. Results convey that the negative impact of the Hindenburg report became pronounced from day 1 of the release of the report. CAR results show that there was a significant decline for all the sample companies during the initial 20 days. After that, the companies partially retained their position. The financial performance analysis further highlights an adverse impact on the companies' quarterly results post-report. The technical analysis revealed significant market reactions following the Hindenburg report, but it's crucial to consider the broader socioeconomic impacts, especially on retail investors in India. The sharp decline in Adani Group's stock prices led to substantial losses for retail investors, who rely on their investments for critical financial goals. Behavioral finance insights show that these investors are particularly prone to overreaction and panic selling, exacerbating market volatility. Additionally, the decline affected numerous suppliers, contractors, and employees linked to the Adani Group, highlighting the interconnectedness of corporate health and broader economic stability. These findings underscore the need for robust market regulations, enhanced transparency, and investor protection measures to safeguard against similar events in the future. The immediate and long-term financial impacts observed in the Adani Group case highlight the need for vigilance and proactive risk management strategies by investors, including portfolio diversification and thorough due diligence on corporate governance practices. Enhanced market transparency and stricter regulatory frameworks are essential to prevent market manipulation and ensure corporate accountability, as suggested by similar studies on market reactions to corporate scandals (Fama, 1970; Karpoff et al., 2008). Regulatory bodies like SEBI must enforce stringent disclosure norms and swiftly address any manipulative practices. By learning from the market's reaction to the Hindenburg report, policymakers can develop strategies to protect investors and maintain market integrity, such as adopting international best practices for market oversight and leveraging technology for better surveillance. These measures can mitigate the risks associated with corporate scandals, as evidenced by studies on the Satyam scandal and the ICICI Bank case (Chaki, 2023). This study underscores the need for robust regulatory frameworks in India to prevent financial misconduct and protect investors. Key gaps include insufficient disclosure requirements, weak enforcement mechanisms, inadequate corporate governance standards, and poor auditor independence. Enhancing transparency in shareholding patterns and improving monitoring and surveillance systems are critical. Addressing these gaps through regulatory reforms will strengthen market integrity and investor confidence, fostering stability and growth in Indian financial markets. Furthermore, this case emphasizes the necessity for corporate governance reforms to prevent future occurrences of similar incidents. Ensuring enhanced board oversight, effective audit committees, and strong whistleblower protections, along with stringent transparency and timely disclosures, is crucial. These reforms are particularly vital in emerging markets like India, where regulatory oversight may not be as rigorous as in more established markets.

References

Adani response (2023). Retrieved from https://www.adani.com/-/media/Project/Adani/Invetsors/Adani-Response-to-Hindenburg-January-29-2023

Agarwal, S., Agarwal, M., & Ghosh, R. (2021). An event study approach to analyse the impact of novel coronavirus disease (Covid-19) on Indian hotel & tourism stocks performance. *Ramanujan International Journal of Business and Research, 6*, 26–36.

Balasubramanian, N., Black, B. S., and Khanna, V. (2010). The relation between firm-level corporate governance and market value: A case study of India. *Emerging Markets Review, 11*(4), 319–340.

Balyuk, T., & Fedyk, A. (2023). Divesting under pressure: US firms' exit in response to Russia's war against Ukraine. *Journal of Comparative Economics, 51*(4), 1253–1273.
Basdas, U., & Oran, A. (2014). Event studies in Turkey. *Borsa Istanbul Review, 14*(3), 167–188.
Bashir, U., Kayani, U. N., Khan, S., Polat, A., Hussain, M., & Aysan, A. F. (2024). Investor sentiment and stock price crash risk: The mediating role of analyst herding. *Computers in Human Behavior Reports, 13*, Article 100371.
Black, B. S., Kim, W., Jang, H., & Park, K. S. (2015). How corporate governance affect firm value? Evidence on a self-dealing channel from a natural experiment in Korea. *Journal of Banking & Finance, 51*, 131–150.
Brown, S. J., & Warner, J. B. (1985). Using daily stock returns: The case of event studies. *Journal of Financial Economics, 14*(1), 3–31.
Business Insider. (2023). *Adani-Hindenburg Saga Timeline: The trouble, the fall, and the turnaround.* Retrieved from https://www.businessinsider.in/business/corporates/news/adani-hindenburg-saga-timeline-the-trouble-the-fall-and-the-turnaround/articleshow/100797291.cms
Chaki, A. B. a. D. (2023). Credit Suisse says won't accept Adani bonds as collateral. *The Economic Times.* Retrieved from https://economictimes.indiatimes.com/markets/stocks/news/credit-suisse-says-wont-accept-adani-bonds-as-collateral/articleshow/97533248.cms
Davidson, W. N., Worrell, D. L., & Lee, C. I. (1994). Stock market reactions to announced corporate illegalities. *Journal of Business Ethics, 13*, 979–987.
Deb, S. G., and Mukherjee, J. (2008). Does stock market development cause economic growth? A time series analysis for Indian economy. *International Research Journal of Finance and Economics, 21*(3), 142–149.
Demirgüneş, K. (2016). The effect of liquidity on financial performance: Evidence from Turkish retail industry. *International journal of economics and finance, 8*(4), 63–79.
Dsouza, J. J. (2016). Quarterly earnings and stock Prices reactions—A study of BSE-500 Companies. *Amity Journal of Finance, 1*(1), 28.
Fama, E. F. (1965). The behavior of stock-market prices. *The Journal of Business, 38*(1), 34–105.
Fama, E. F. (1970). Efficient capital markets. *Journal of Finance, 25*(2), 383–417.
Ghosh, A. (2006). Determination of executive compensation in an emerging economy. Evidence from India. *Emerging Markets Finance and Trade, 42*(3), 66–90.
Hindenburg Research. (2023). *Adani Group: How the world's 3rd richest man is pulling the largest con in corporate history.* Retrieved from https://hindenburgresearch.com/adani/
Hu, D. (2014). Japanese stock market reaction to announcements of news affecting auditors' reputation: The case of the Olympus fraud. *Journal of Contemporary Accounting & Economics, 10*(3), 206–224.
Irungu, A. M., Muturi, W., Nasieku, T., & Ngumi, P. (2018). ISSN NO: 2616–4965.
Jain, P. (2022). Regulatory actions against corporate irregularities in India: analyzing the stock market impact. *Cogent Economics & Finance, 10*(1), 2122187.
Jayshree P Upadhyay (2023). Indian regulator's probe faults Adani group on disclosure rules - sources. Retrieved from Indian regulator's probe faults Adani group on disclosure rules -sources | Reuters
Karpoff, J. M., Lee, D. S., & Martin, G. S. (2008). The cost to firms of cooking the books. *Journal of Financial and Quantitative Analysis, 43*(3), 581–611.
Kolaric, S., and Schiereck, D. (2016). Are stock markets efficient in the face of fear? Evidence from the terrorist attacks in Paris and Brussels. *Finance Research Letters, 18*, 306–310.
Kothari, S. P., & Warner, J. B. (2007). Econometrics of event studies. In *Handbook of empirical corporate finance* (pp. 3–36). Elsevier.
İltaş, Y., and Demirgüneş, K. (2020). Asset tangibility and financial performance: A time series evidence. *Ahi Evran Üniversitesi Sosyal Bilimler Enstitüsü Dergisi, 6*(2), 345–364.
MacKinlay, A. C. (1997). Event studies in economics and finance. *Journal of Economic Literature, 35*(1), 13–39.

Mallesha, L., & Archana, H. N. (2023). Impact of Hindenburg research report on the stock prices of Adani group companies: An event study. *Asia-Pacific Journal of Management Research and Innovation, 19*(1), 40–46.

Pritchard, A. C., & Ferris, S. P. (2001). *Stock price reactions to securities fraud class actions under the private securities litigation reform act.* Available at SSRN 288216.

Rahman, M. L., Amin, A., & Al Mamun, M. A. (2021). The COVID-19 outbreak and stock market reactions: Evidence from Australia. *Finance Research Letters, 38*, Article 101832.

Song, C., & Han, S. H. (2017). Stock market reaction to corporate crime: Evidence from South Korea. *Journal of Business Ethics, 143*, 323–351.

Varma, J., R. (1997). Corporate governance in India: Disciplining the dominant shareholder. http://hdl.handle.net/11718/9522

Wang, T. Y., Winton, A., & Yu, X. (2010). Corporate fraud and business conditions: Evidence from IPOs. *The Journal of Finance, 65*(6), 2255–2292.

Zhu, Y., Wu, Z., Zhang, H., and Yu, J. (2017). Media sentiment, institutional investors and probability of stock price crash: evidence from Chinese stock markets. *Accounting & Finance, 57*(5), 1635–1670.

Assessing the Impact of ESG Initiatives on Financial Performance in *the* Energy Sector from Developing *Countries*: Efficiency Analysis and Productivity Growth by Data *Envelopment* Analysis and Malmquist Index

Sumita Biswas, Rashmi Singh, and Niladri Das

Abstract In the last few decades, Environmental Social Governance (ESG) has experienced significant growth in industry and corporations. This research evaluates the energy efficiency and productivity growth in energy firms from developing countries using data from the Bloomberg terminal from 2016–2023. The study uses data envelop analysis (DEA) and the malmquist index to assess the performance of energy firms. Results show significant year-to-year variations in efficiency, highlighting the influence of ESG factors on performance outcomes. Solar energy, oil, and gas firms demonstrate robust operational and financial performance, reflecting their adaptability to regulatory pressures and technological advancements. The paper concludes by suggesting the continuous integration of ESG strategies and technology innovations as a route to sustainable efficiency and competitiveness in the new global energy environment.

Keywords ESG · Financial Performance · Data Envelop Analysis · Corporations

S. Biswas · R. Singh (✉) · N. Das
Department of Management Studies & Industrial Engineering, Indian Institute of Technology (ISM) Dhanbad, Dhanbad, India
e-mail: rashmis@iitism.ac.in

S. Biswas
e-mail: 23dr0331@iitism.ac.in

N. Das
e-mail: niladri@iitism.ac.in

B. Basu et al. (eds.), *Data Analytics for Decision Making towards Business Excellence*, Uncertainty and Operations Research, https://doi.org/10.1007/978-981-95-2429-7_10

1 Introduction

Environmental, Social, and Governance (ESG) has gained global recognition due to global challenges such as climate change-related natural disasters and the global financial crisis (Alam et al., 2024). In today's economy, investing in green or socially responsible projects has grown much more prevalent (Egorova et al., 2022). Sustainable development is regarded as one of the significant worldwide trends influencing the growth of contemporary businesses (Egorova et al., 2022). The company's environmental actions focus on promoting environmental sustainability by following regulations. Its social actions involve handling stakeholders and communities effectively. Governance ensures integrity and ethical behaviour within management systems, while ESG encompasses corporate management tasks that comply with legal requirements, moral principles, and environmental and social accountability (Chang & Lee, 2022).

In the early 1970s, researchers searched for a connection between ESG elements and corporate financial performance (CFP) (Friede et al., 2015). Since then, more than 2500 empirical studies and multiple review studies on this relationship have been published by researchers and investors (Friede et al., 2015). The company's environmental, social, and governance actions are governed by legal requirements, moral principles, and environmental and social accountability, ensuring the company's ethical and responsible business practices (Kim & Li, 2021). ESG factors have been found to positively impact equity prices and profitability, but they also play a crucial role in addressing uncertainties and mitigating financial risks (Derwall et al., 2005). ESG factors significantly impact corporate credit risks, influencing credit ratings and strengthening the connection between ESG practices and financial performance (Kim & Li, 2021). First, ESG practices in governance establish robust decision-making and compliance frameworks, reducing mismanagement and regulatory breaches, thus enhancing risk management (Kim & Li, 2021). Second, companies with proactive ESG strategies demonstrate higher resilience to challenges, leading to improved creditworthiness, reduced borrowing costs, and greater access to capital (Kim & Li, 2021). Additionally, A robust ESG profile reassures investors, stabilizes equity demand, and enhances market valuation, while mitigating non-financial risks like climate change, social unrest, and governance failures (Liu et al., 2022). Lastly, Meeting stakeholder expectations enhances reputation and minimizes risks. Integrating ESG factors into core strategies improves financial outcomes and builds long-term resilience, solidifying the connection between ESG practices and financial performance (Liu et al., 2022; Nirino et al., 2021).

The Sustainable Development Goals (SDGs) emphasize the role of technological innovation in addressing global challenges, especially in emerging economies where pro-growth policies often compromise environmental quality (Chishti & Sinha, 2022).Regional policies and government interventions are crucial in addressing resource distribution and inequality, but economic shocks and global market dynamics also impact these efforts (Van Der Linden & Freeman, 2017). Technology has potential to tackle climate change and support SDGs, but evidence on structured policies is limited, necessitating cohesive strategies for sustainable and inclusive development (Chishti & Sinha, 2022; Sinha et al., 2020; Van Der Linden & Freeman, 2017).

Previous research evident that capital markets aligned with sustainable development play a crucial role in bridging the financing gap for achieving the Sustainable Development Goals (SDGs) (Zhan & Santos-Paulino, 2021). Over the past decade, there has been a significant increase in sustainability-focused financial products, with stock exchanges serving as platforms and the UN's 2030 Agenda for Sustainable Development providing a globally recognized framework (Zhan & Santos-Paulino, 2021). Increasingly, companies and institutional investors are recognizing the importance of aligning their investment and business decisions with positive development outcomes (Zhan & Santos-Paulino, 2021). Economic growth determines a country's international market position, while sustainable development ensures its longevity across nations (Siakwah et al., 2020).

Research indicates that developed nations prioritize socio-cultural rights, environmental and economic sustainability more than developing nations (Yoon et al., 2018). The developed and developing world have ESG practices and reporting structures, but gaps exist in legislation, penalties, and incentives for further ESG push (Yoon et al., 2018). The business environment in developing and emerging economies differs significantly from developed nations, with companies facing different regulatory environments, social pressures, and less early planning and reporting of their corporate social responsibility activities (Naeem & Çankaya, n.d.).

The Kuznets curve reveals that countries' ESG disclosure levels vary based on economic development stages, with distinct relationships between environmental pollution and economic production (Van Der Linden & Freeman, 2017). The evolution of ESG frameworks in various nations is highlighted in Table 1, providing a snapshot of their legal frameworks and initiatives aimed at fostering sustainable practices. These differences underscore the need for a tailored approach to ESG implementation, taking into account the unique challenges and opportunities that each country faces based on its level of development.

Policymakers and regulators are prioritizing corporate environmental governance and social responsibility, leading to increased research on ESG factors and their impact on corporate value (Cojoianu et al., 2021), social responsibility performance (Albuquerque et al., 2019; Mishra & Modi, 2013), and corporate governance performance (Drakos & Bekiris, 2010; Petersen, 2009) positively impact corporate market value. These findings indicate that improving ESG performance increases corporate market value (Zhou et al., 2022). Corporate strategies shift from profit maximization to social interest maximization, focusing on environmental protection, social responsibility, and governance, driven by investor concerns about ESG-based investments (Iazzolino et al., 2023). The extensive study on ESG operations and strategies' impact on corporate financial performance is extensive, but further research is needed to understand its impact on ecologically conscious businesses in developing nations (Garcia et al., 2017). Research shows the energy sector faces severe environmental sustainability concerns (Vázquez & Iribarren, 2015). Due to their significant environmental impact, these companies invest heavily in environmental management to

Table 1 ESG frameworks placed in developing nations (Singhania et al., 2024)

ESG framework	Country	Legal framework	Description
ESG framework at an early stage	India	Guidance note on nonfinancial Disclosure, 2011	To motivate companies to voluntarily expand their disclosures beyond a narrow focus on financial reporting
		Department of Public enterprises guideline on CSR and sustainability for CPSE, 2013	Setting objectives for a CPSE to allocate a fixed percentage towards CSR and sustainability initiatives
		NVG on Social, Environmental and Economic Responsibilities of Bussiness Revolutionaries	Integrating CSR as a core component of overall business policies
ESG framework at early stage	Nigeria	Code of CG for Public companies of Nigeria, 2008	Issued the 2003 Code of Corporate Governance and implemented sustainability disclosure
	Vietnam	Sustainability Reporting handbook for Vietnamese Companies, 2013	To give recommendations on integrating sustainability in organizations
ESG framework at early stage	South Africa	King Code Report on Corporate Governance, 1994, 2002 and 2010	Requires integrated reporting and third party assurance of reports
		JSE social responsible Index (SRI), 2004	Encourages companies listed on the FTSE/JSE All Share Index to report on ESG pillars
ESG framework at developing stage	Malaysia	Malaysian Code for Institutional Investors, 2014	A framework to facilitate the funding of sustainable and responsible investment initiatives
		Listing Requirements for Sustainability Statement, 2015	Including sustainability statements in annual reports addressing key sustainability issues
ESG framework at early stage	China	Guidelines for Social Responsibility in Outbound Mining Investments	Adopting sustainability reporting as a voluntary practice aligned with international standards
		Guidelines on Insurance Industry Social Responsibility	To promote the adoption of social responsibility within the insurance industry

address the issues arising from their operations (Miralles et al., 2018). In recent years, the energy sector in developing countries has faced many challenges, such as resource management, changes in energy demand and consumption patterns, uncertainty regarding new energy projects, and lower investment (Siksnelyte, 2021). They are experiencing unprecedented energy demand due to rapid population growth rates and the pursuit of economic development (Jiang & Martek, 2021). This study highlights a critical gap in energy efficiency research. The focus largely been on technical and financial aspects. However, environmental and social dimensions remain overlooked particularly in developing countries (Popkova & Sergi, 2021). ESG criteria integration is crucial for sustainable development and energy sector financial performance, navigating regulatory complexities, attracting sustainable investments, and strengthening long-term resilience. Promoting economic initiatives aligning with environmental and social goals fosters global collaboration for energy efficiency enhancement (Popkova & Sergi, 2021). This study further examines the energy sector's efficiency and productivity in developing nations. It focuses on the relationship between ESG performance and financial returns. The aim is to enhance resource utilization, operational performance, risk management, regulatory compliance, and stakeholder relationships.

This paper is structured into four sections. The second section provides a literature review on ESG and corporate performance and theoretical framework. In section three, the data, and research methodology is discussed in detail. Section four provides results and discussion and later on conclusion, and future direction is discussed.

2 Literature Review

The literature review is divided into two sections: the first focuses on the review of ESG, while the second provides a theoretical framework.

2.1 *Brief Review on ESG and Corporate Performance*

Several studies, both local and international, demonstrated that the ESG standard can raise the short run costs but enhance the company's performance on an operational level in the medium and long runs (Dhar et al., 2022). Prior research also indicates that business investments in environmental protection and social responsibility lower listed enterprises' equity of spending and their economic performance (Dhar et al., 2022). However, according to the "social impact hypothesis," ESG inputs can improve corporate financial performance (CFP) (Chen et al., 2022). The theory suggests that fulfilling social responsibilities can enhance organizational impact, staff sense of belonging, and performance by prioritizing a supportive workplace, implementing talent development programs, empowering employees, and boosting

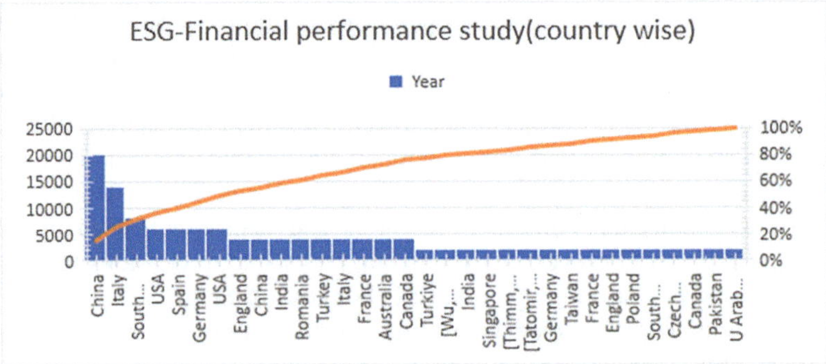

Fig. 1 ESG-Financial Performance study 2016–2023 (country-wise) (Author's Compilation)

productivity through comprehensive initiatives (Chen et al., 2022). If businesses overshoot in ESG, it reduces their financial performance Chen et al., 2022; Grassmann, 2021). It's been evident that ESG is a crucial tool for achieving carbon neutrality, with countries like China aiming for it by 2060. Canada, South Africa, Japan, and South Korea are also working towards this goal. By mid-century, two-thirds of global emissions and 75% of GDP are expected to be decarbonized (H.-M. Chen et al., 2022). The KPMG 2020 Global CEO Foresight Survey indicates that 63% of leaders will prioritize ESG in the volatile global environment since the COVID-19 pandemic.(S. Chen et al., 2023). ESG investments, which focus on environmental and social aspects, offer strict risk control measures to mitigate environmental and social risks, increasing investor interest. Corporate concern for social responsibility and environmental protection is often considered a "gold brand" (S. Chen et al., 2023). Companies that do better in this area tend to have better growth potential. So they can also fit in with asset managers' quest for long-term stable returns (H.-M. Chen et al., 2022).

Meta-studies over several decades have explored the connection between ESG and financial performance (Bergeri et al., 2022). Most analyses, published before 2015, found positive links between ESG performance, operational efficiencies, and stock performance. Studies between 2015 and 2020 reviewed 1500 corporate cases, revealing that 58% assessed ESG-financial performance relationships using indicators like ROE, ROA, ROC, or stock price. Among these, 13% found a neutral impact, 21% showed mixed results, and only 8% reported a negative relationship. Country-wise research, shown in Fig. 1, highlights significant contributions from China, Italy, South Africa, and the USA, advancing understanding in this field.

2.1.1 Theoretical Framework

The ESG matter in the literature concerning firm value gets support from the stakeholder theory as well as the resource-based view theory (RBV) literature (Behl et al., 2022). Both theories support ESG practices' impact on firm value (Behl et al., 2022). According to earlier research, integrated corporate social responsibility (CSR) and environmental, social, and governance (ESG) activities result in the creation of special intangible resources, such as expertise in developing better, less polluting technologies (Behl et al., 2022). These practices help to build corporate culture through a proactive stance on social disclosures and reputation built because of better integration with the environment and social practices as well as governance requirements (Behl et al., 2022).According to the stakeholder theory (Freeman, 2010), the company is supposed to maximize wealth for all stakeholder groups and not just shareholders. Accordingly, the activities of the business firm are expected to lead to accounting-related as well as social and envirLuionmental outcomes. Hence, the company needs both financial and non-financial performance metrics to assess the value created for all its stakeholders. Diversity and inclusion policies aim to meet the needs of multiple interested parties (Suárez-Fernández et al., 2024). Thus, a company's ESG strategy shall be better supported by its stakeholders in developed countries compared with stakeholders in developing countries, which shall take its toll on the outcomes of the ESG implementation (Ruan & Liu, 2022). The relationship between ESG and CFP is industry and sector-specific because of the differences between industries and sectors (Liu et al., 2022).

3 Data and Methodology

The study used the DEA model to assess the efficiency of 100 energy firms in developing countries, using the Malmquist Productivity Index to measure energy productivity performance. The output variables were Return on Equity, Return on Assets, and Return on Capital, while the input variables were Environment, Social, and Governance. Table 2 details input and output variables based on prior study expertise and data availability. ESG criteria are used as input variables and financial performance metrics as output variables. This helps firms and policymakers make informed decisions focusing on Environmental, Social, and Governance (ESG) can benefit energy companies, particularly those in developing nations, by enhancing stakeholder relations, operational efficiency, reducing environmental impact, and strengthening governance. The financial health and success of these companies can be gauged through ROE, ROA, and ROC. Given the growth of ESG sentiment, firms beating on an increasing number of areas could be assumed to become more attractive investments (Shiller, 2000). Data is considered over the period from 2016 to 2023 of energy sector in developing nations (Fig. 2). R Studio and DEAP 2.1 software are used for the analysis part. Details of the study are represented in Table 3. The reasons for choosing this study period are as follows: First, 2016–2017 marks a significant global awareness and adaptation of ESG standards. With this energy, companies in developing nations have started synchronizing with global standards to draw in foreign investments and enhance their market standing (Hoang, 2018). The details of the study are attached in Appendix 1.

Table 2 Description of input and output variables. (*Author's Compilation*)

Input and output variables	Secondary indicator	Description	Reference
ESG	Environmental	Environmental factors can significantly impact a sovereign's, entity's, or person's financial performance or solvency	(Damodaran, 2007; Li et al., 2021; Orlitzky et al., 2003)
	Social	Social matters could influence an entity, sovereign, or individual's financial performance or solvency in a favorable or undesirable way	
	Governance	Governance matters significantly impact the financial performance or solvency of any entity, sovereign, or individual	
Financial performance	Return on equity	Return on equity (ROE) is the ratio of a company's net income to its shareholders' equity	
	Return on assets	Market analysts widely use return on assets (ROA) to measure firm performance, as it measures the effectiveness of assets in producing income	
	Return on Capital	ROC is the ratio of a company's net income to its debt and equity value, indicating the company's earnings above its average cost of debt and equity capital	

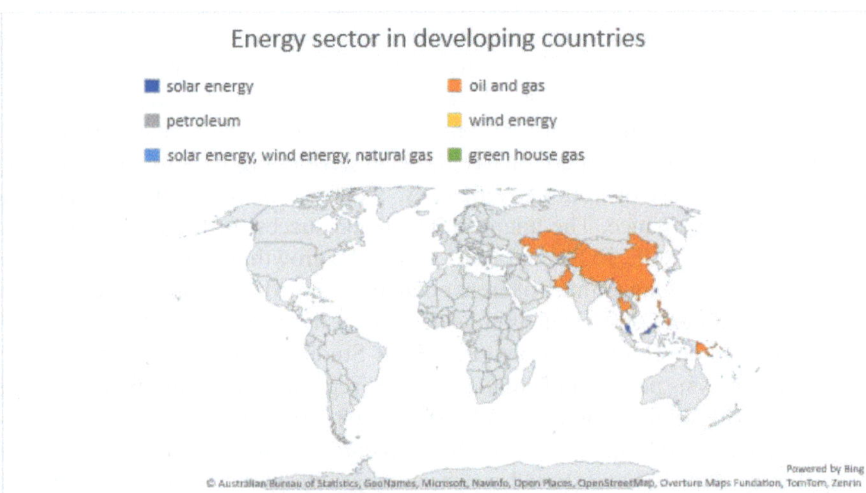

Fig. 2 Energy firms of developing countries (country-wise) (Author's Compilation)

Table 3 Correlation coefficient between variables (*Author's Compilation*)

Variables	Environment	Social	Governance	Return to equity	Return to assets	Return to capital
Environment	1					
Social	–0.03803	1				
Governance	–0.14476	–0.37989	1			
Return to equity	0.534494	–0.29716	–0.06112	1		
Return to assets	–0.01084	0.632166	–0.35772	0.308008	1	
Return to capital	0.572	0.027407	–0.08339	0.389027	0.333456	1

During 2018–2019, there was a rapid increase in the adaptation of the ESG framework (Hoang, 2018). Simultaneously there was a stronger push for transparency and accountability in corporate governance 2020–2021 marked a profound impact on the energy sector due to COVID-19. This period saw significant disruptions in supply chains and shifting towards energy demands. Evaluating data from these years helps understand how energy firms managed these unprecedented challenges and the role of ESG in their recovery strategies (Gollakota & Shu, 2023; Kuzemko et al., 2020). 2022–2023, the post-pandemic period, shows an acceleration toward green energy transitions. Over this period, significant technological advancements in the energy sector have been made, including developing renewable energy technologies, smart grids, and energy storage solutions (Buonomano et al., 2023). The necessary empirical data for the research in this paper are collected from the Bloomberg database. The mean is employed to ensure the robustness and reliability of the research findings in handling the missing data. Additionally, by applying Min–Max normalization techniques, we ensured comparability across different scales, thereby enhancing the validity of the analysis. The correlation matrix between input and output variables is presented in Table 4.

This facilitates comprehension of the direction and magnitude of the linear relationship between the input and output variables.

3.1 Correlation Matrix Between Input and Output Variables

Table 3 indicates that there is a moderate positive correlation (0.534) between ROE and Environment, suggesting that higher Environmental Impact scores are associated with higher ROE. Social factors show a moderate positive correlation (0.632) with ROA, indicating that higher Social Responsibility scores correlate with higher ROA. Moreover, ROC displays a moderate positive correlation (0.572) with the environment and a more negligible positive correlation (0.389) with ROA, suggesting that both Environmental Impact and ROA are moderately related to ROC. Hence, considering the moderate correlations determined between ESG factors and financial performance metrics in the above tables, the choice of the Charnes, Cooper, and Rhodes DEA model over any other model is justified. According to Dellnitz et al. (2018), this model allows us to measure the efficiency associated with converting ESG efforts into financial returns while assuming constant returns to scale (CRS).

3.1.1 DEA-CCR Model

The CCR DEA model was first introduced by Charnes, Cooper, and Rhodes in 1978 assuming constant returns to scale (Charnes et al., 1978). It is assumed in CRS that all DMUs will work on an optimal scale. This efficiency for each DMU is defined as the ratio of the weighted output summation to the weighted input summation. Firms' DEA efficiency scores often fall between zero and one. On the other hand, a firm that scores "one" on the efficient frontier curve is considered relatively efficient; on the other hand, a firm that scores less than "one" is considered relatively inefficient. To measure the relative efficiency of any DMUp, the traditional DEA model is constructed as a linear divisive programming problem

$$\text{Maximize}\, Zpq = \frac{\left(\sum_{i=1}^{e} UipXiq\right)}{\left(\sum_{j=1}^{f} VjpYjq\right)} \tag{1}$$

Subject to

$$q = 1, 2, 3 \ldots\ldots n \tag{2}$$

$$0 \leq \frac{\left(\sum_{i=1}^{e} UipYiq\right)}{\left(\sum_{j=1}^{f} VjpXjq\right)} \leq 1\, q = 1, \ldots\ldots n \tag{3}$$

$$Uip \geq € \cdots i = 1, \ldots e \tag{4}$$

$$Vip \geq €j = 1 \ldots, f \tag{5}$$

where

$n =$ number of DMUs units to be evaluated.

$e =$ number of outputs.

$f =$ number of inputs.

Uip Weighted attached to i-th output ($i = 1$.........e) for DMUp ($p = 1$............ n).

Vjp Weighted attached to j-th input ($j = 1$........f) for DMUp ($p = 1$............ n)

Xiq value of *i*– th output (*i*= 1……e) for DMU*p* (*p* = 1………… *n*)

Yjq value of *j*– th output (*j*= 1……f) for DMU*p* (*p* = 1………… *n*)

Zqp relative efficiencies of DMU *q* (*q*= 1………. *n*) when evaluated using the weighted associated with DMU*p* (*p* = 1………. *n*).

€ =infinitesimal constant.

Equation (1) maximizes the efficiency of the considered DMU*p*. Equation (2) defines the efficiencies of all DMUs*q* concerning the weights chosen for *p*, while Equation (3) ensures that all efficiencies are between 0 and 1. Equations (4) and (5) ensure all weights of outputs and inputs are positive.

DEA-CCR Model Formulation:

For each DMUpq (where q = 1,2,...,100), the efficiency score is given by the ratio of the weighted sum of outputs to the weighted sum of inputs.

$$\text{Maximize}\, Zpq = \frac{\left(\sum_{i=1}^{3} UipXiq\right)}{\left(\sum_{j=1}^{3} VjpYjq\right)} \tag{6}$$

Subject to

$$q = 1, 2, 3 \ldots\ldots 100 \tag{7}$$

$$0 \leq \frac{\left(\sum_{i=1}^{3} UipXiq\right)}{\left(\sum_{j=1}^{3} VjpYjq\right)} \leq 1\, q = 1, \ldots\ldots 100 \tag{8}$$

$$Uip \geq € \cdots i = 1, 2, 3 \tag{9}$$

$$Vjp \geq € j = 1, 2, 3 \tag{10}$$

where

n = number of DMU units to be evaluated.

e = number of output variables.

f = number of inputs variables.

Uip Weighted attached to *i*- th output (*i* = 1,2, 3 corresponding to ROE, ROA, and ROC) for DMU*p* (*p* = 1………… 100).

Vjp Weighted attached to *j*- the input (*j* = 1,2, 3 corresponding to Environment, Social and Governance) for DMU*p* (*p* = 1…… 100)

$$Xiq\, value\, of\;\, i - th\, output\; (i = 1, 2, 3)\, for\; \text{DMU}\, p\; (p = 1 \ldots\ldots 100)$$

$$Yjq\, value\, of\; j - th\, output\; (i = 1, 2, 3)\, for\; \text{DMU}\, p\; (p = 1 \ldots\ldots 100)$$

Zpq = relative efficiencies of DMU q (q= 1.......... 100) when evaluated using the weighted associated with DMUp (p = 1........... 100).

€ =infinitesimal constant.

Here, Equation (6) maximizes the efficiency of the considered DMUp. Equation (7) defines the efficiencies of all hundred DMUsq for weights chosen for p, while Equation (8) ensures that all efficiencies are between 0 and 1. Equations (9) and (10) ensure all weights of outputs and inputs are positive. The detailed study framework of the research is illustrated in Fig. 3.

Understanding productivity changes in the energy sector is crucial, involving examining efficiency change, which measures resource utilization, and technological change, which indicates technological advancements contributing to productivity growth. The Malmquist Productivity Index is used to quantify these changes from 2016 to 2023 for a hundred Decision-Making Units (DMUs), providing a comprehensive view of productivity trends in the energy sector.

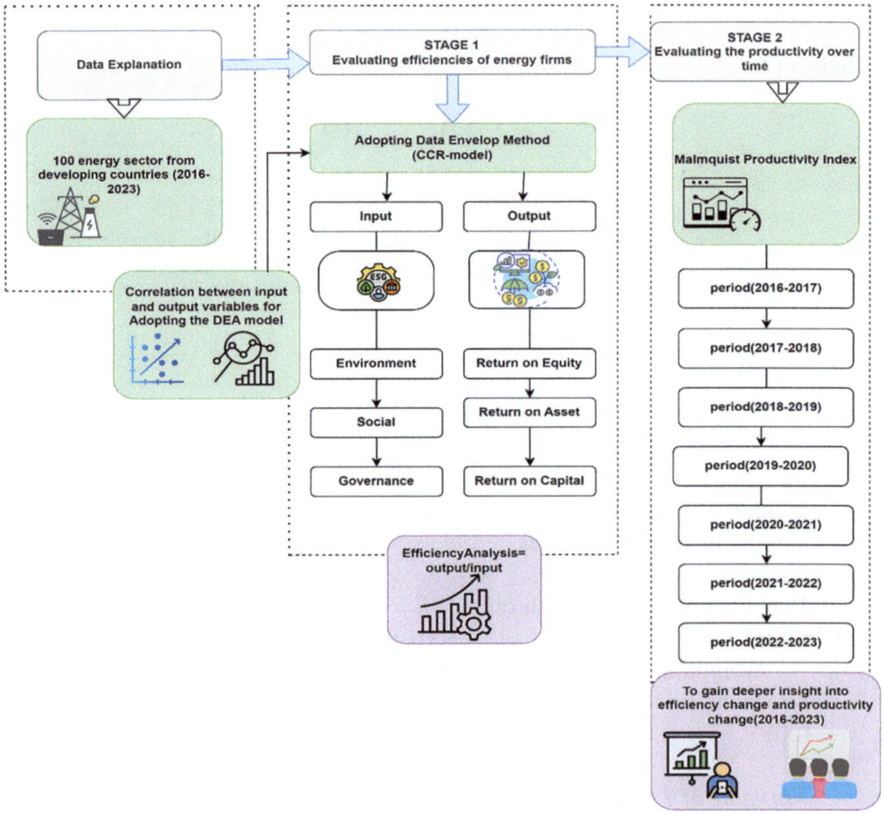

Fig. 3 The study framework (Author's Compilation)

3.1.2 Malmquist Index

The non-parametric Malmquist productivity index is a formal time series analysis method for comparing the efficiency of DMUs across time intervals by figuring out several DEA-style models The study evaluates efficiency in three inputs and outputs, revealing that productivity of a DMU improves with an MPI greater than one, remains unchanged with an equal MPI, and decreases with an MPI less than one (Al-Refaie et al., 2016). This research aims to assess the productivity changes of the hundred energy firms sector from 2016–2023. Besides it evaluates the increase in efficiency and productivity growth between seven-year energy strategic plans during 2016–2017, 2017–2018, 2019–2020, 2020–2021, 2021–2022, and 2022–2023.

4 Result and Discussion

In the present study, the efficiency of one hundred energy firms in developing countries was analyzed using the DEA-CCR model, which accounts for undesirable output. Moreover, the Malmquist productivity index exhibiting undesirable results has been used to trace the total factor productivity change throughout the study period. Table 4 shows the efficiency analysis of a hundred DMU units from 2016–2023.

The result indicates energy firms specializing in solar energy, oil and gas, and petroleum exhibited perfect efficiency (1.00) for several years. These energy firms have headquarters in China and Malaysia. Some headquarters of these energy companies are based in China and Malaysia. In recent years, Malaysia and China are among the countries trying to foster sustainable development. This has been manifested through various activities as well as policies enacted by both nations' governments (Yatim et al., 2016). It has also been evident that Chinese and Malaysian companies have had considerable developments in the PV industry in recent years (Liu et al., 2022; Zhang et al., 2020). Firms are embracing environmentally sustainable energy production methods, supported by favourable tax and regulatory policies, leading to improved financial performance. The Renewable Energy Transition Roadmap 2035 (RETR 2035), proposed in Malaysia's 12th Plan, aims to enhance system stability while aligning with environmental targets and strengthening the nation's economy. This initiative is strategically linked with financial performance metrics and Environmental, Social, and Governance (ESG) criteria, ensuring Malaysia's energy sector evolves sustainably while contributing positively to economic growth (Chachuli et al., 2021).

GCL SYSTEM INTEGRATION TEC-A, a China-based industry, has shown low efficiency in recent years due to rapid technological advancements in solar PV cells, creating a competitive post-pandemic environment. However, another solar energy firm named JA-SOLAR TECHNOLOGY CO LTD-A, a solar energy firm, demonstrated a significant increase in efficiency from 0.176 in 2016 to 0.715 in 2023. On the other hand, Thailand-based firm BCPG PCL-FOREIGN achieved perfect efficiency scores of 1.000 in 2020 and 2021 but experienced a notable drop to 0.133 in 2023. This

Table 4 Efficiency analysis of energy firm (2016–2023) (*Author's Compilation*)

DMU	2016	2017	2018	2019	2020	2021	2022	2023
DMU 1	1	0.227	0.389	0.139	0.227	0.354	0.261	0.234
DMU 2	0.808	0.690	1	0.532	1	0.623	0.770	0.460
DMU 3	0.688	0.517	0.840	0.398	0.703	0.339	0.338	0.304
DMU 4	0.452	0.546	0.816	0.492	0.703	0.339	0.338	0.304
DMU 5	1	0.397	0.504	0.418	0.703	0.339	0.338	0.304
DMU 6	0.816	0.343	0.183	0.090	0.020	0.374	0.487	0.477
DMU 7	0.816	0.343	0.183	0.090	0.029	0.374	0.487	0.477
DMU 8	1	0.343	0.183	0.063	0.029	0.374	0.487	0.477
DMU 9	1	0.264	0.342	0.299	0.365	0.506	0.437	0.444
DMU 10	1	0.245	0.407	0.123	0.370	0.215	0.580	0.464
DMU 11	0.691	0.258	0.394	0.124	0.267	0.390	0.394	0.133
DMU 12	0.796	0.278	0.399	0.118	0.301	0.390	0.394	0.133
DMU 13	1.000	0.206	0.357	0.103	0.270	1	0.580	0.464
DMU 14	0.513	0.260	0.327	0.164	0.340	0.506	0.437	0.444
DMU 15	0.175	0.222	0.319	0.124	1	1	0.394	0.133
DMU 16	0.028	0.042	0.130	0.047	0.022	1	0.158	0.191
DMU 17	0.103	0.074	0.152	0.042	0.037	1	0.158	0.191
DMU 18	0.192	0.170	0.336	0.227	0.219	0.220	0.493	0.451
DMU 19	0.192	0.170	0.336	0.227	0.219	0.220	0.493	0.451
DMU 20	0.558	0.185	0.336	0.172	0.219	0.220	0.493	0.530
DMU 21	0.883	0.012	0.437	0.284	0.007	0.536	1	0.472
DMU 22	0.148	0.030	0.237	0.099	0.023	0.151	0.421	0.291
DMU 23	0.098	0.410	0.457	0.017	0.023	0.362	1	0.437
DMU 24	0.176	0.086	0.927	0.870	0.646	0.595	0.739	0.715
DMU 25	0.456	0.147	0.927	0.870	0.646	0.639	0.739	0.696
DMU 26	0.535	0.134	0.144	0.133	0.366	0.434	0.457	0.646
DMU 27	0.529	0.259	0.230	0.072	0.366	0.420	0.457	0.634
DMU 28	0.219	0.322	0.321	0.764	0.395	0.291	0.075	0.072
DMU 29	0.467	0.418	0.278	0.144	0.302	0.296	0.075	0.198
DMU 30	0.644	0.574	0.269	0.083	0.024	0.242	0.445	0.304
DMU 31	0.644	0.490	0.269	0.083	0.024	0.235	0.445	0.304
DMU 32	0.615	0.802	0.269	0.116	0.024	0.235	0.445	0.359
DMU 33	1	0.911	1	1	1	0.458	0.781	0.727
DMU 34	0.711	0.654	1	1	1	0.475	0.781	0.977
DMU 35	0.567	0.295	0.473	0.202	0.527	0.466	1	1
DMU 36	0.335	0.267	0.356	0.187	0.502	0.465	1	1
DMU 37	0.480	0.202	0.263	0.168	0.190	0.320	1	0.432

(continued)

Table 4 (continued)

DMU	2016	2017	2018	2019	2020	2021	2022	2023
DMU 38	0.449	1	0.418	0.244	1	0.427	0.240	0.316
DMU 39	0.354	1	0.279	0.381	0.706	0.432	0.240	0.316
DMU 40	0.474	1	0.403	0.341	0.986	0.432	0.240	0.399
DMU 41	0.477	0.940	0.696	0.797	1	0.103	0.773	0.919
DMU 42	0.527	1	0.723	1	0.951	0.932	1	0.595
DMU 43	0.569	0.730	0.723	1	0.951	0.932	1	0.580
DMU 44	0.861	0.821	0.431	0.200	0.201	0.418	0.260	0.320
DMU 45	0.861	0.821	0.431	0.200	0.270	0.418	0.260	0.320
DMU 46	0.812	0.620	0.431	0.574	0.270	0.418	0.260	0.310
DMU 47	0.820	0.816	0.227	0.509	0.819	0.528	0.543	0.284
DMU 48	0.884	0.342	0.227	0.577	0.819	0.537	0.274	0.495
DMU 49	1	1	0.628	1	1	1	0.763	0.867
DMU 50	1	0.787	0.628	1	1	1	0.763	1
DMU 51	0.292	0.270	0.114	0.003	0.074	0.298	0.454	1
DMU 52	1	0.443	0.169	0.350	0.251	0.013	0.153	0.534
DMU 53	1	0.586	0.298	0.129	0.251	0.012	0.153	0.534
DMU 54	0.267	0.453	0.162	0.086	0.181	1	0.153	0.030
DMU 55	0.370	0.516	0.201	0.139	0.244	0.425	0.153	0.238
DMU 56	0.820	0.989	0.477	0.225	0.528	0.443	0.131	0.262
DMU 57	0.998	1	0.641	0.108	0.430	0.512	0.131	0.233
DMU 58	0.555	1	0.286	0.015	0.394	0.155	0.018	0.611
DMU 59	0.264	0.240	0.136	0.089	0.233	0.443	0.255	0.974
DMU 60	0.242	0.351	0.079	0.565	0.306	0.434	0.255	0.931
DMU 61	0.668	0.542	0.868	0.665	0.800	0.432	0.204	0.545
DMU 62	0.960	0.452	0.868	0.157	0.800	0.185	0.204	0.540
DMU 63	0.488	0.822	0.717	0.248	0.417	0.221	0.577	0.498
DMU 64	0.365	0.558	1	0.504	0.534	0.030	0.577	1
DMU 65	0.787	0.010	0.518	0.442	0.196	0.635	0.259	0.513
DMU 66	0.489	0.073	0.227	0.174	0.269	0.047	0.103	0.297
DMU 67	0.687	0.464	0.502	0.115	0.269	0.130	0.103	0.273
DMU 68	0.604	0.471	0.234	0.006	0.023	0.639	0.096	0.029
DMU 69	0.302	0.268	0.760	0.004	0.021	1	0.017	0.031
DMU 70	0.303	0.686	0.760	0.004	0.413	0.321	0.017	0.031
DMU 71	0.692	0.556	0.333	0.003	0.402	0.321	0.017	0.518
DMU 72	0.285	0.098	0.601	0.095	0.939	1	1	0.741
DMU 73	0.048	0.205	0.489	0.178	0.775	1	1	0.695
DMU 74	0.232	0.326	0.353	0.140	0.208	0.236	0.214	0.476

(continued)

Table 4 (continued)

DMU	2016	2017	2018	2019	2020	2021	2022	2023
DMU 75	0.245	0.395	0.159	0.136	0.112	0.207	0.214	0.400
DMU 76	0.030	0.028	0.206	0.075	0.202	0.405	0.279	0.209
DMU 77	0.036	1	0.315	0.276	0.049	0.273	0.119	0.327
DMU 78	0.185	1	0.015	0.334	0.040	0.460	0.119	0.327
DMU 79	1	0.549	0.256	0.160	0.733	0.321	0.074	0.413
DMU 80	1	0.336	0.592	0.365	1	1	0.304	0.695
DMU 81	1	0.129	0.592	0.309	1	0.597	0.249	0.750
DMU 82	0.062	0.641	0.334	0.001	0.168	0.128	0.010	1
DMU 83	0.043	0.197	0.475	0.319	0.217	1	1	1
DMU 84	0.466	0.328	0.611	0.866	0.251	0.277	0.392	1
DMU 85	0.230	0.400	0.245	0.576	0.640	0.295	0.235	1
DMU 86	0.200	0.184	0.370	0.208	0.421	0.328	0.235	1
DMU 87	0.546	0.163	0.572	0.180	0.282	0.313	0.095	0.608
DMU 88	0.930	0.014	0.704	0.097	0.282	0.340	0.095	1
DMU 89	0.343	0.327	0.091	0.053	0.058	0.199	0.083	0.157
DMU 90	0.217	1	0.096	0.012	0.103	0.063	0.036	0.306
DMU 91	0.608	0.306	0.481	0.005	0.812	0.015	0.067	1
DMU 92	0.659	0.196	0.227	0.406	0.412	0.139	0.862	1
DMU 93	0.535	0.036	0.291	0.282	0.563	0.104	0.418	1
DMU 94	0.472	0.051	0.687	0.420	0.468	0.821	0.418	1
DMU 95	0.267	1	0.207	0.232	0.083	0.411	0.051	0.721
DMU 96	0.673	1	0.016	0.234	0.534	0.106	0.064	0.182
DMU 97	0.356	0.122	1	0.876	1	1	0.613	0.222
DMU 98	0.354	0.198	1	0.270	0.995	1	0.436	0.445
DMU 99	0.321	0.198	0.263	0.207	0.373	0.190	0.308	1
DMU 100	0.216	0.198	1	0.207	0.373	0.191	0.308	1

fluctuation could be attributed to changes in market demand for renewable energy and pricing dynamics within the energy sector. Malaysian oil and gas company Yinson Holdings Berhad consistently maintains high efficiency, starting with a perfect score in 2016 and maintaining relatively high scores throughout the period. Scores range from 0.139 in 2019 to 0.389 in 2018, indicating some fluctuation but generally strong performance. Thus, we can conclude that the firm's commitment to high efficiency in energy operations aligns with its ESG goals, driving sustainable business practices and strong financial performance over the years (Khalil et al., 2023). SUZHOU MAXWELL TECHNOLOGIE-A, a solar energy firm, achieves perfect efficiency scores (1.000) in 2016, 2018, and 2019. However, a slight decrease is observed in the latter years, ending with 0.727 in 2023. This pattern is consistent among many solar energy firms based in China, Malaysia, and Thailand, demonstrating their ability to maintain high-efficiency scores over multiple years. An analysis is conducted on the distribution of inefficient energy firms during 2016–2023. Figure 4 shows the ratio of efficient and inefficient from 2016 to 2023 and the distribution of inefficient DMUs. The analysis of inefficient DMUs will offer valuable insights into potential interventions to improve energy sector efficiency and productivity. Table 5 explains the year-wise trend and variation in performance from Fig. 4.

The bar charts in Fig. 5 display the number of times each Decision-Making Unit (DMU) appears in reference sets, comparing them to other DMUs. From 2016 to 2023, Solar Energy firms consistently appear as efficient benchmarks across multiple charts. This suggests their stable and high performance in the renewable energy sector. The frequency of appearances varies significantly among other types of energy firms. The Oil and Gas Firm is a reliable benchmark in the energy sector, appearing 69 times, but less frequently than top-performing solar energy firms. Solar Energy Firms are efficient units, serving as references for developing countries, while Oil and Gas Firms play a significant role. The study reveals that solar energy companies are exhibiting exceptional operational performance, setting industry standards in efficiency. It is essential to create a benchmark for energy firms' efficiency across different countries. By comparing performance metrics, stakeholders can identify leading countries and understand factors contributing to their efficiency, which can inform best practices and policy development. Figure 6 provides a comprehensive view of the energy sector's efficiency landscape among developing countries, offering country-wise efficiency analysis from 2016 to 2023. To deepen the understanding of efficiency ratings, we further analyze the unique challenges faced by different types of energy sectors.

Solar energy: From 2016 to 2023 (Table 4), most firms, especially those in China and Malaysia, saw increased efficiency due to the sustainability of solar energy. Solar energy is a secure and reliable renewable energy source, producing minimal pollution and reducing global warming emissions (Shahsavari & Akbari, 2018). It also has the potential to significantly reduce energy-related greenhouse gas emissions, helping to mitigate climate change effects. Solar energy also provides communities with stability against fossil fuel price volatility (Shahsavari & Akbari, 2018). The advancements in solar energy technologies, such as the applications of solar photovoltaic (PV) systems, off-grid electrification, and solar water heating, have made it increasingly cost-competitive with traditional energy sources, as noted by Painuly and Wohlgemuth (2001). These factors collectively contribute to the consistent efficiency and growing adoption of solar energy over the years.

Oil and gas: Apart from solar energy, the oil and gas sector is demonstrating high energy efficiency, especially in developing countries like China, Malaysia, and Thailand. Population growth drives oil demand growth, boosting energy consumption. Countries like China and Australia have strong prospects for enhancing gas production capacity, with China planning to increase gas production by 2.5 times by 2035 (Ediger & Berk, 2018). Oil consumption assessments experienced significant variations between 2012 and 2015 (Table 4) due to differing economic evaluations

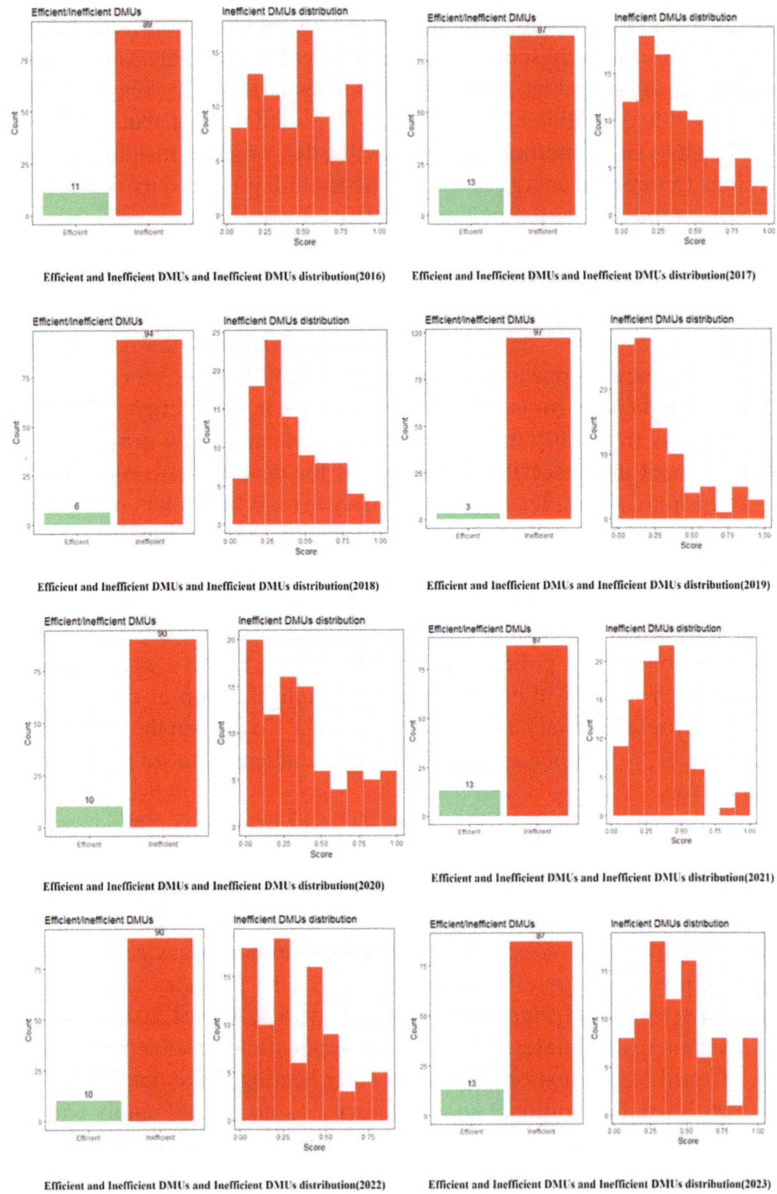

Fig. 4 Efficient and inefficient DMU distribution (2016–2023) (Author's Compilation)

Table 5 Year-wise comprehensive analysis of inefficient DMUs. (*Author's Compilation*)

Year	Comprehensive analysis
2016	There were eleven efficient DMUs. Approximately thirteen inefficient DMUs had efficiency scores of less than 0.25, indicating significant underperformance. Over 30 DMUs scored below 0.50, showing a substantial portion of inefficiency. Additionally, nearly 12 DMUs were close to achieving efficiency, with scores ranging from 0.75 to 1.0. This distribution suggests a polarized performance, with many DMUs performing poorly and a few nearing efficiencies
2017	In 2017, 13 efficient DMUs were found, while 87 were found inefficient. The dispersion of efficiency scores increased, with visible bunching within the 0.2 to 0.05 range. This suggests a wider range of inefficiency across many DMUs, making it challenging to achieve higher efficiency
2018	In 2018, six efficient DMUs were identified, with a peak around 0.3 and evenly distributed scores across the spectrum. Many DMUs were clustered around moderate inefficiency, with few at very low or high levels
2019	In 2019, efficiencies were concentrated between 0.1 and 0.3, with many DMUs performing poorly, with a significant proportion remaining at very low levels
2020	In 2020, efficiency scores spread wider, with peaks around 0.3 and 0.5, indicating varying performance levels among DMUs, with some improving and others remaining inefficient
2021	The efficiency distribution in 2021 closely resembled 2017, with a peak of approximately 0.5, indicating a fair scatter between 0.1 and 0.7 scores, indicating diversity in efficiency levels across the involved DMUs
2022	In 2022, there are ten efficient DMU units, with most inefficient DMUs having scores between 0.25 and 0.50, with a significant number having very low scores close to zero
2023	In 2023, the density of efficient DMUs was higher around 0.5, with a flat distribution across other ranges, and the number of efficient DMUs was consistently low, indicating only a small fraction achieved perfect efficiency

Summary The efficiency scores of inefficient DMUs consistently fall below 0.5, indicating varying challenges and opportunities based on their ESG practices and financial strategies. The dynamic balance of environmental, social, governance, and financial considerations suggests that integrating these areas can improve efficiency and overall performance for DMUs.

following the global financial and economic crisis, despite consistent energy efficiency over the years (Morozov et al., 2018). The oil and gas sector faces challenges, particularly in developing nations, including rising regulatory pressures, fluctuating global oil prices, and outdated infrastructure (Ediger & Berk, 2018). Technological barriers, limited access to advanced solutions, environmental risks, social and political pressures, and the global shift towards renewable energy pose challenges to efficiency and environmental sustainability (Morozov et al., 2018).

Wind energy: Wind energy, while efficient in many developing countries, lacks the efficiency of solar or oil and gas due to limitations. Despite its sustainability, wind turbines face challenges like lack of expertise in site selection and technical aspects, operational failures, and poorly screened projects. These issues limit the overall efficiency and adoption of wind energy compared to solar and oil and gas energy (Abramowski and Posorski, 2000).

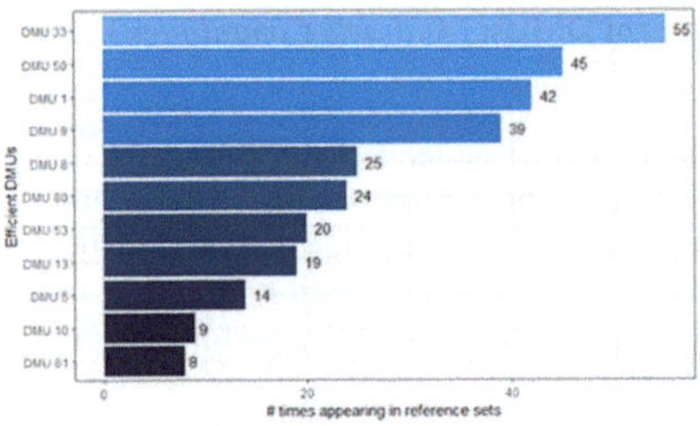

Efficient DMUs Vs times appearing in reference sets(2016)

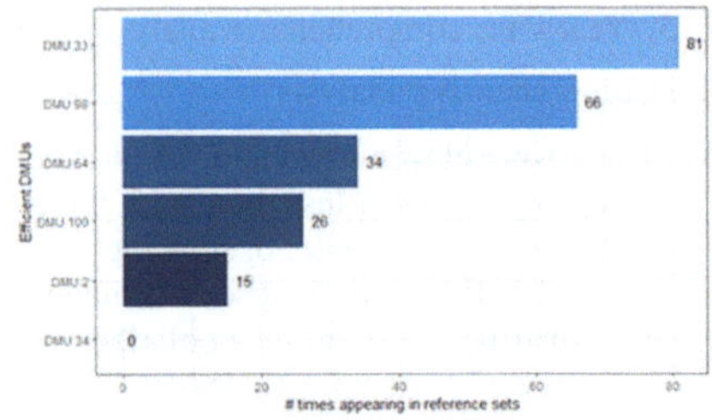

Efficient DMUs Vs times appearing in reference sets(2017)

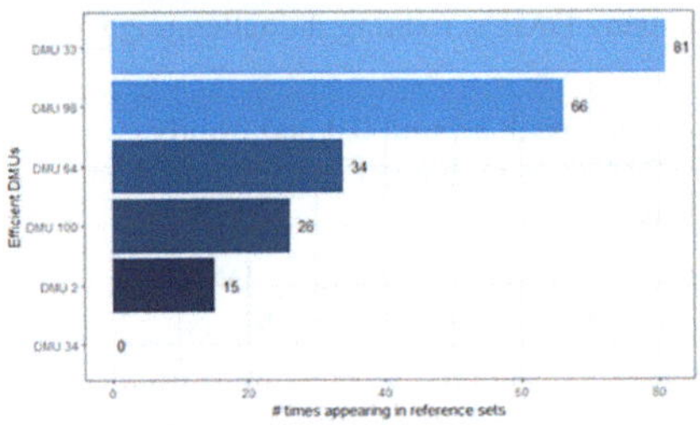

Efficient DMUs Vs times appearing in reference sets(2018)

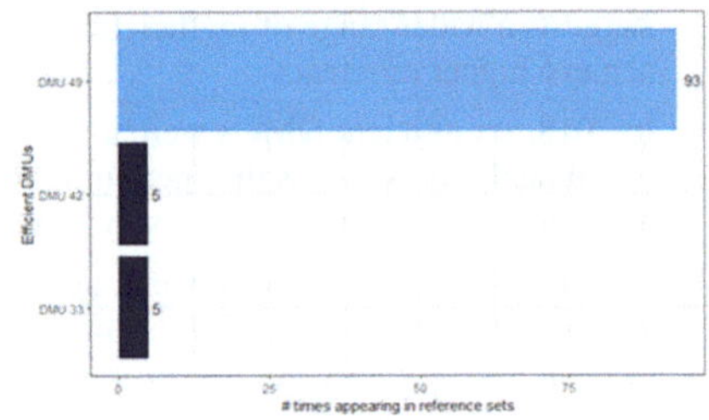

Efficient DMUs Vs times appearing in reference sets(2019)

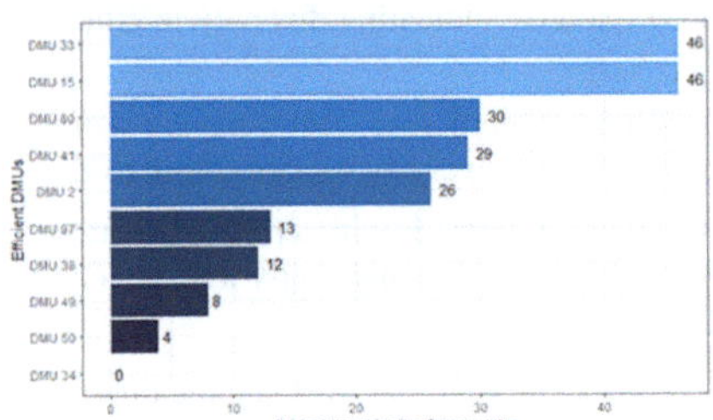

Efficient DMUs Vs times appearing in reference sets(2020)

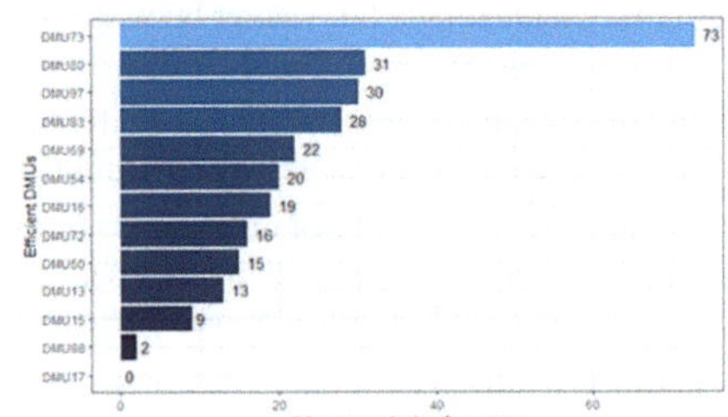

Efficient DMUs Vs times appearing in reference sets(2021)

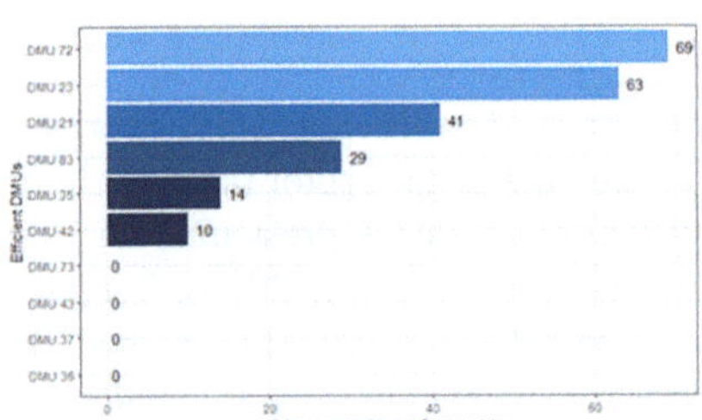

Efficient DMUs Vs times appearing in reference sets(2022)

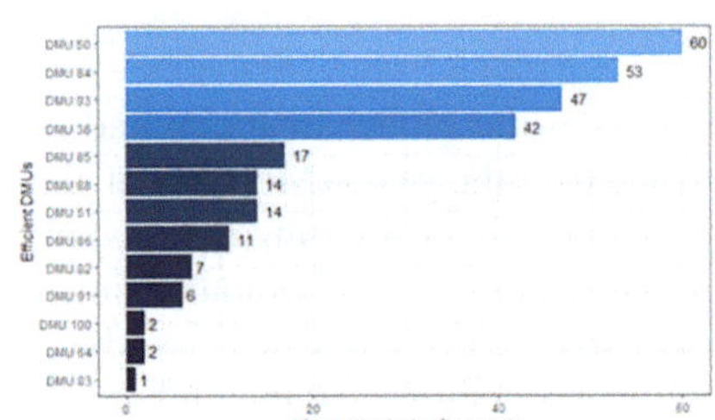

Efficient DMUs Vs times appearing in reference sets(2023)

Fig. 5 Efficient DMUs Vs times appearing in reference sets (2016–2023) (Author's Compilation)

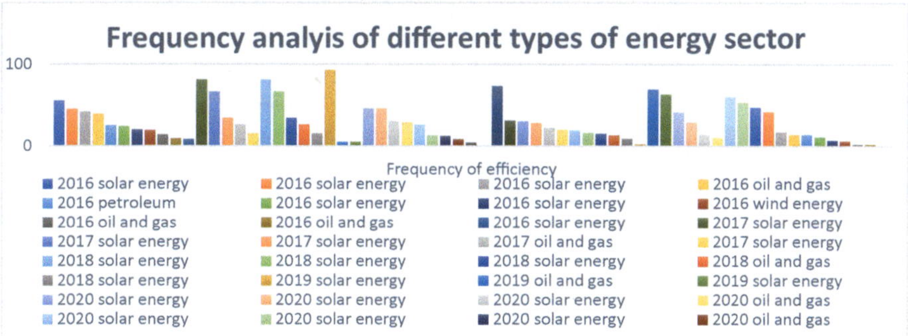

Fig. 6 Energy frequency analysis of different types of the energy sector (2016–2023) (Author's Compilation)

The chart (Fig. 6) shows the frequency analysis of efficiency across various energy sectors from 2016 to 2020. Solar energy shows the highest frequency, indicating significant technological advancements and widespread adoption, particularly in years like 2017 and 2020. Oil and gas maintain substantial frequency, despite sustainability challenges and fluctuating demand. Wind energy shows lower frequency, suggesting it may not have achieved the same level of efficiency improvements. Petroleum shows minimal frequency, indicating its declining role in the global energy mix. The data highlights solar energy's growing dominance in efficiency advancements, while oil and gas remain relevant.

To gain deeper insights, a country-wise analysis was conducted (Fig. 7). The results reveal that China consistently demonstrated higher efficiency performance from 2016 to 2023. Malaysia recorded the highest efficiency in 2016, while Kazakhstan achieved peak efficiency between 2022 and 2023. Taiwan reached its highest efficiency in 2018, and Pakistan exhibited an upward trend in energy sector efficiency in 2022. Assessing the productivity changes of these energy firms over this period is crucial for understanding the dynamics of productivity growth and identifying long-term trends and patterns.

4.1 *Malmquist Result*

The Malmquist Productivity Index is a tool that provides a comparative analysis of changes in the productivity of Direct Market Units (DMUs) over time, revealing the influence of Environmental, Social, and Governance (ESG) factors on these energy firms(Woo et al., 2015)This analysis helps policymakers and industry stakeholders evaluate the effectiveness of policies promoting ESG integration and sustainability practices. The data highlights the growing influence of ESG factors on efficiency outcomes, especially in 2021–2022 following the COVID-19 pandemic (Table 6).

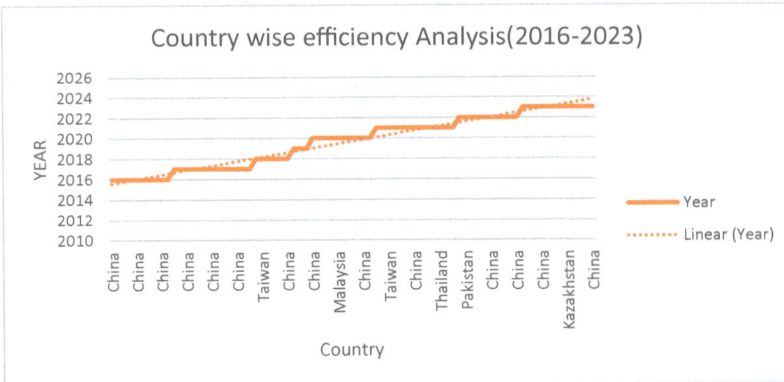

Fig. 7 Country-wise efficiency performance (2016–2023) (Author's Compilation)

Furthermore, in recent years, scholarly research on how ESG affects a company's market value and financial performance has expanded quickly (Chen et al., 2022). With a mean score of 1.120, the total factor productivity change shows notable fluctuations throughout the periods. This suggests that total factor productivity has increased, primarily due to technical developments rather than purely efficiency gains. Efficiency gains are important, but ongoing investments in technological innovation and sustainable practices are necessary for long-term productivity and financial performance increases. Technological developments improve financial performance in addition to increasing efficiency.

This study uses Tobin's Q ratio to evaluate the financial performance and growth potential of firms across various sectors, particularly in energy, oil, gas, and renewable industries. A higher Q value indicates strong market confidence and growth prospects, often linked to operational efficiency or innovation, as seen in clean energy firms. Conversely, lower Q values may signal inefficiency or limited market optimism. By comparing Q values across firms and regions, this study aims to identify trends in market valuation, assess investor perceptions, and understand the implications for sustainable development and energy transitions.

Tobin q ratio: TQ is the ratio between a physical asset's market value and its replacement value. The market value of a company's assets is measured by the market value of its outstanding stock and debt, whilst the replacement cost of assets is measured using their book value. A ratio of 1 or more indicates that the firm's market value exceeds that of its recorded assets(Singh et al., 2011). Thus,

$$\text{TQ ratio} = \frac{\textit{total market value of company} + \textit{liabilities}}{\textit{total assets value} + \textit{liabilities}}$$

Table 7 presents a list of companies with their respective Tobin's Q values, and financial indicators evaluating a firm's market performance based on asset replacement costs. A Q value greater than 1 indicates high market value, indicating growth potential, while a Q value lower than 1 suggests inefficiency. Sectors include energy,

Table 6 Malmquist index summary of annual means (2016–2023) (*Author's Compilation)*

Year wise	Efficiency change	Technical change	Pure technical efficiency change	Scale efficiency change	Total factor productivity change	Comment
2016–2017	0.923	1.124	0.939	0.983	1.079	For energy companies, both overall efficiency and pure technical efficiency somewhat decreased. It is, however, mostly countered by observable technology advancements that caused a minor decline in scale efficiency
2017–2018	0.629	2.043	0.667	0.943	1.285	Technological advancements have led to a positive productivity change despite a decline in efficiency
2018–2019	1.127	1.826	1.414	0.902	2.375	High levels of technology and efficiency changes led to substantial productivity gains
2019–2020	0.464	2.442	0.823	0.563	1.123	The results highlight the dominant role of technology in driving productivity

(continued)

Table 6 (continued)

Year wise	Efficiency change	Technical change	Pure technical efficiency change	Scale efficiency change	Total factor productivity change	Comment
2020–2021	1.661	0.388	0.865	1.919	0.644	Efficiency surged high but a low technological change showing the limitations of efficiency improvements without technological
2021–2022	0.898	2.409	0.948	0.948	2.164	Efficiency changes are highly emphasizing impact of ESG
2022–2023	1.112	0.372	1.075	1.034	0.414	Efficiency gains alone were insufficient for significant productivity improvement
mean	0.923	1.214	0.939	0.983	1.120	Technological advancements have significantly impacted total factor productivity change, with notable improvements observed in specific periods

oil, gas, and solar. This study analyzes Tobin's Q values, key financial metrics used to assess a firm's market performance relative to the replacement cost of its assets, across various sectors, including energy, oil, gas, and renewable energy. High Tobin's Q values, such as GCL System Integration Technology (7.99) and Suzhou Maxwell Technology (4.49), indicate strong market confidence and perceived growth potential. These high values are particularly prominent in the clean energy sector, where companies like LONGI-Green Energy Technology and Tongwei Co Ltd-A also display

Table 7 Tobin q ratio of the firm *(Author's compilation)*

Firm name	Tobin q value	Firm name	Tobin q value
YINSON HOLDINGS BHD (DMU 1)	0.82	TRINA SOLAR CO LTD-A (DMU 26)	1.93
YINSON HOLDINGS BHD-CW25 (DMU 2)	1.00	TRINA SOLAR CO LTD-A (DMU 27)	1.93
PTT OIL & RETAIL BUS-FOREIGN (DMU3)	1.05	ORIENTAL ENERGY CO LTD -A (DMU 28)	1.00
PTT OIL & RETAIL BUSINESS PC (DMU 4)	2.71	ORIENTAL ENERGY CO LTD -A (DMU 29)	1.00
PTT OIL & RETAIL BUSINE-NVDR (DMU 5)	2.71	THAI OIL PCL-FOREIGN (DMU 30)	0.53
BANGCHAK CORP PCL-FOREIGN (DMU6)	2.71	THAI OIL PCL (DMU 31)	0.53
BANGCHAK CORP PCL (DMU7)	0.82	THAI OIL PCL-NVDR (DMU 32)	0.53
BANGCHAK CORP PCL-NVDR (DMU8)	0.82	SUZHOU MAXWELL TECHNOLOGIE-A (DMU 33)	4.49
NORINCO INTL COOPERATION -A (DMU9)	1.47	SUZHOU MAXWELL TECHNOLOGIE-A (DMU 34)	4.49
SICHUAN NEW ENERGY POWER C-A (DMU10)	2.02	HENGDIAN GROUP DMEGC -A (DMU 35)	2.43
BCPG PCL-NVDR (DMU 11)	0.40	HENGDIAN GROUP DMEGC -A (DMU 36)	2.43
BCPG PCL (DMU12)	0.40	EXXON MOBIL CORP (DMU 37)	1.97
SICHUAN NEW ENERGY POWER C-A (DMU 13)	2.03	PTG ENERGY PCL-NVDR (DMU 38)	4.68
NORINCO INTL COOPERATION -A (DMU 14)	1.13	PTG ENERGY PCL (DMU 39)	4.68
BCPG PCL-FOREIGN (DMU 15)	0.59	PTG ENERGY PCL-FOREIGN (DMU 40)	4.68
GCL SYSTEM INTEGRATION TEC-A (DMU 16)	7.99	PETRONAS GAS BHD (DMU 41)	2.54
GCL SYSTEM INTEGRATION TEC-A (DMU 17)	7.99	LONGI-GREEN ENERGY TECHNOL-A (DMU 42)	2.32
PTT EXPLOR & PROD PCL-NVDR (DMU 18)	0.98	LONGI-GREEN ENERGY TECHNOL-A (DMU 43)	2.32

(continued)

Table 7 (continued)

Firm name	Tobin q value	Firm name	Tobin q value
PTT EXPLOR & PROD PUBLIC CO (DMU 19)	0.98	PTT PCL/FOREIGN (DMU 44)	0.82
PTT EXPLORATION & PROD-FOR (DMU 20)	0.99	PTT PCL (DMU 45)	0.82
CONOCOPHILLIPS (DMU 21)	2.57	PTT PCL-NVDRN (DMU 46)	0.82
SANTOS LTD (DMU 22)	0.98	TCL ZHONGHUAN RENEWABLE EN-A (DMU 47)	2.58
OCCIDENTAL PETROLEUM CORP (DMU 23)	1.81	TCL ZHONGHUAN RENEWABLE EN-A (DMU 48)	2.58
JA-SOLAR TECHNOLOGY CO LTD-A (DMU 24)ara>	1.88	TONGWEI CO LTD-A (DMU 49)	3.01
JA-SOLAR TECHNOLOGY CO LTD-A (DMU 25)	1.88	TONGWEI CO LTD-A (DMU 50)	3.01
ATTOCK REFINERY LTD (DMU 51)	0.146	BUMI ARMADA BERHAD (DMU 76)	1.46
RISEN ENERGY CO LTD-A (DMU 52)	1.05	SINOPEC OILFIELD SERVICE C-A (DMU 77)	1.24
RISEN ENERGY CO LTD-A (DMU 53)	1.05	SINOPEC OILFIELD SERVICE C-A (DMU 78)	1.24
COFCO BIOTECHNOLOGY CL LT-A(DMU 54)	1.02	SAN MIGUEL CORP (DMU 79)	0.017
COFCO BIOTECHNOLOGY CL LT-A (DMU 55)	1.02	FLAT GLASS GROUP CO LTD-A (DMU 80)	1.63
CHINA PETROLEUM & CHEMICAL-A (DMU 56)	0.92	FLAT GLASS GROUP CO LTD-A (DMU 81)	1.63
CHINA PETROLEUM & CHEMICAL-A (DMU 57)	0.92	NOSTRUM OIL & GAS PLC (DMU82)	1.31
HENGYUAN REFINING CO BHD (DMU 58)	0.55	CNOOC LTD-A (DMU83)	1.21
JINKO SOLAR CO LTD-A (DMU 59)	2.88	CNOOC LTD-A (DMU 84)	1.21
JINKO SOLAR CO LTD-A (DMU 60)	2.88	MING YANG SMART ENERGY GRO-A (DMU 85)	1.02
ZHEJIANG CHINT ELECTRICS-A (DMU 61)	1.2	MING YANG SMART ENERGY GRO-A (DMU 86)	1.02

(continued)

Table 7 (continued)

Firm name	Tobin q value	Firm name	Tobin q value
ZHEJIANG CHINT ELECTRICS-A (DMU 62)	1.2	GOLDWIND SCIENCE & TECHNOL-A (DMU 87)	1.06
SHANGHAI AIKO SOLAR ENERGY-A (DMU 63)	3.32	GOLDWIND SCIENCE & TECHNOL-A9 (DMU 88)	1.06
SHANGHAI AIKO SOLAR ENERGY-A (DMU 64)	3.32	PETRON CORP (DMU 89)	0.20
PETRONAS DAGANGAN BHD (DMU 65)	3.22	SHANGHAI AEROSPACE AUTOMOB-A (DMU 90)	2.43
CHINA OILFIELD SERVICES-A (DMU 66)	1.29	SHANGHAI AEROSPACE AUTOMOB-A (DMU 91)	2.43
CHINA OILFIELD SERVICES-A (DMU 67)	1.29	PRIMA MARINE PCL-NVDR (DMU 92)	1.86
MALAYSIA MARINE AND HEAVY EN (DMU 68)	0.44	PRIMA MARINE PCL (DMU 93)	1.86
IRPC PCL – NVDR (DMU 69)	1.07	PRIMA MARINE PCL-FOREIGN (DMU 94)	1.86
IRPC PCL – FOREIGN (DMU 70)	1.07	SINOPEC OILFIELD EQUIPMENT-A (DMU 95)	2.03
IRPC PCL (DMU 71)	1.07	SINOPEC OILFIELD EQUIPMENT-A (DMU 96)	2.03
XINJIANG DAQO NEW ENERGY C-A (DMU 72)	1.01	HOYUAN GREEN ENERGY CO LTD-A (DMU 97)	0.89
XINJIANG DAQO NEW ENERGY C-A (DMU 73)	1.01	HOYUAN GREEN ENERGY CO LTD-A (DMU 98)	0.89
PETROCHINA CO LTD-A (DMU 74)	1.06	CNOOC ENERGY TECHNOLOGY & -A (DMU 99)	1.66
PETROCHINA CO LTD-A (DMU 75)	1.06	CNOOC ENERGY TECHNOLOGY & -A (DMU 100)	1.66

robust Q values, signaling investor optimism toward renewable energy and sustainability initiatives. In contrast, firms with lower Q values, such as San Miguel Corp (0.017) and Hengyuan Refining (0.55), may face concerns about operational efficiency or limited growth prospects. Similarly, companies like BCPG PCL (0.40) and PTT Exploration & Production (0.98) reflect modest market expectations, often tied to inefficiencies or challenges in asset utilization. Solar and renewable energy firms generally stand out with higher Q values, driven by global transitions toward clean energy and sustainability. The study reveals that developed companies like ExxonMobil and Sinopec maintain stable Q values, indicating market confidence, while emerging markets like GCL System Integration Technology and LONGI-Green Energy show higher Q values, indicating the growing significance of renewable energy in global investment trends.

5 Conclusion

The study emphasized the utilitarian use of the CCR-DEA method to measure and assess the information on the efficiency of energy firms in developing countries. The obtained results have not been compared with the results obtained by other authors due to their innovative nature and the lack of research results predisposed to discussion. In this research, the DEA-CCR model is used to analyze the efficiency of energy firms. Additionally, the Malmquist index is employed to assess the energy efficiency and productivity growth of the energy sector in developing countries from 2016 to 2023. Two cases are evaluated: the energy efficiency country-wise over the period and the total productivity change from 2016–2023 (Fig. 6). Secondly, productivity changes during this period (2016–2023). Three input factors (ESG) and three output factors (financial performance) are identified for our study. Research shows that sustainable and responsible investments are now major concerns for investors and creditors, recognizing the importance of environmental sustainability and socio-economic stability. Corporations worldwide focus more on long-term ESG goals than profit maximization, prioritizing ESG strategies for their transparency and sustainability. According to many findings, ESG performance is correlated in a significant way with the financial performance of energy and power generation corporations (Naeem & Çankaya, n.d.). Correlation analysis is carried out to examine the factor relations. Technical efficiency and pure technical efficiency are computed using the CCR model. Then, TIE, PTIE, and SIE, the technical, pure technical, and scale inefficiency averages, are determined, in same order. Efficiency has varied significantly year-wise, with some years showing improvements (e.g., 2016–2017, 2018–2019) and others showing declines (e.g., 2017–2018, 2019–2020). The efficiency changes in 2021–2022 emphasize the impact of ESG factors, suggesting that these considerations can influence efficiency outcomes (Chen et al., 2022). Technological advancements have consistently been the primary driver of total factor productivity change.

While efficiency improvements are beneficial, they are often only sufficient with concurrent technological progress. The data underscores the critical role of technology in driving productivity gains over the years reviewed. Financially, firms that consistently appear as benchmarks, such as Solar Energy firms and occasionally Oil and Gas firms, demonstrate strong operational and financial performance. The post-pandemic situation has led to opportunities for structural changes and the development of jobs with low productivity and little added value in the modern economy. Solar PV is considered the cheapest option for generating electricity in much of the world (Wang et al., 2022). Developing countries, in particular, are increasingly turning to solar PV as a pivotal strategy for achieving economic growth that is both affordable and sustainable. This trend highlights their efficiency in adopting renewable energy solutions. It underscores their capacity to adapt to technological advancements and regulatory demands, including those concerning Environmental, Social, and Governance (ESG) criteria in developing countries. The future direction for energy sectors, including solar energy and oil and gas, will continue to be shaped by technological advancements, the integration of ESG factors, and the need to adapt to an evolving regulatory landscape. Solar energy firms are poised to maintain their leadership in efficiency, driven by ongoing technological innovation, sustainability commitments, and policy support. However, other sectors, like oil and gas, face the challenge of transitioning towards greater efficiency and sustainability to remain competitive amid growing environmental concerns and shifting consumer expectations.

Beyond energy firms, other industries in developing and emerging economies may also face uncertainties related to their environmental impact, resource management, and integration of ESG principles into business strategies. For instance, industries such as manufacturing, agriculture, and transportation will need to evaluate their sustainability practices, considering factors like supply chain emissions, waste management, and resource use efficiency. These sectors may encounter uncertainties such as fluctuating market demand, changes in consumer preferences, the implementation of stricter environmental regulations, and the potential for technological disruption. Evaluating the integration of ESG considerations in these sectors will help ensure that companies are not only meeting regulatory requirements but also remaining resilient and competitive in the face of global environmental and social challenges.

5.1 Limitation

The present study primarily relies on the DEA-CCR model and the Malmquist index to evaluate the efficiency and productivity of energy firms. These models are well-established in operational research and are suitable for analyzing complex input–output relationships without requiring a predefined functional form. This methodological choice aligns with the study's focus on recent performance trends (2016–2023) and emphasizes the growing influence of ESG factors on firm efficiency, particularly in developing countries. However, there are notable limitations. The DEA-CCR and Malmquist index may not capture all dimensions of efficiency, especially qualitative factors such as policy impacts or technological advancements. Moreover, data reliability in developing countries can be inconsistent, potentially affecting the robustness of findings. The limited sample size and focus on specific energy sub-sectors restrict the study's ability to generalize results across broader regions or industries. The absence of comparative analysis with similar studies further limits the validation and broader applicability of findings. Additionally, the short-term study period may not fully account for long-term trends or cyclical variations in energy efficiency and productivity growth. While ESG factors are acknowledged, their explicit integration into the efficiency and productivity frameworks remains limited. Future research could address these gaps by exploring alternative DEA methods, incorporating comparative analyses, extending the timeline, and explicitly integrating ESG metrics into efficiency models for a more comprehensive understanding of energy firm performance.

5.2 Implication for Academics

The current service makes multiple contributions to the expanding corpus of research. Only a little research has been carried out on the efficiency of the energy sector in developing countries post-pandemic research. In the past, the impact of ESG performance on energy corporations, particularly during times of crisis, has not received much attention. The environment, social responsibility, and good corporate governance can all be used to increase ROA, ROE, and ROC, the study's findings indicate. The data shows that different countries in developing nations contribute to ESG in varying amounts. The evidence from recent studies on the connection between ESG and financial performance needs to be more extensive and precise. Previous studies provide generally consistent results at the overall ESG level. The research identifies a gap in the literature by highlighting the innovative nature of the obtained results and the need for comparative studies. This calls for further comparative research to validate findings and expand understanding of energy efficiency and productivity growth.

5.3 Implication for Policymakers

The research highlights the financial, environmental, and socio-economic benefits of incorporating Environmental, Social, and Governance (ESG) principles into corporate strategies, particularly in the oil and gas sector. Despite the lack of direct results, ESG promotion significantly advances sustainability initiatives, ensuring businesses operate sustainably, enhance reputation, gain stakeholder trust, and achieve Sustainable Development Goals.

The study suggests that policymakers can integrate ESG criteria into corporate strategies, particularly in the energy sector, by promoting sustainable investments and responsible business practices. This approach can unite national and global sustainability objectives, reflect technological advancements, and integrate with broader ESG efforts. Policies that encourage business transparency, accountability, and sustainability can incentivize environmental sustainability and socio-economic stability, leading to better economies. The study also shows that, despite different national features, oil and gas-producing nations in China and Thailand struggle to increase energy efficiency over time. Using this knowledge, policymakers can promote expenditures in R&D and technology to increase the oil and gas industry's energy efficiency. Countries may solve the inefficiencies in their energy sectors and help lower CO2 emissions and energy consumption by promoting technical innovation.

The study emphasizes the importance of supporting renewable energy projects to reduce dependence on fossil fuels and improve energy efficiency. It suggests advocating for sustainable power generation, aligning utilities with decarbonizing efforts, and promoting energy access and security in the Global South. Governments can implement energy conservation programs, such as public awareness campaigns and energy audits, to improve energy efficiency. Modernizing energy infrastructure and investing in advanced grid technologies can also contribute to this modernization.

Declaration: No AI/Gen AI assistance has been taken for generating the chapter.

Appendix

Table A Summary of the study (*Author's Compilation*)

Entity name	Location	Exchange: Ticker	Entity name	Location	Exchange: ticker
YINSON HOLDINGS BHD (DMU 1)	Malaysia	YNS MK Equity	ATTOCK REFINERY LTD (DMU 51)	Pakistan	ATRL PA Equity
YINSON HOLDINGS BHD-CW25 (DMU 2)	Malaysia	7293WA MK Equity	RISEN ENERGY CO LTD-A (DMU 52)	China	300,118 C2 Equity
PTT OIL & RETAIL BUS-FOREIGN (DMU3)	Thailand	OR/F TB Equity	RISEN ENERGY CO LTD-A (DMU 53)	China	300,118 CH Equity
PTT OIL & RETAIL BUSINESS PC (DMU 4)	Thailand	OR TB Equity	COFCO BIOTECHNOLOGY CL LT-A (DMU 54)	China	000,930 CH Equity
PTT OIL & RETAIL BUSINE-NVDR (DMU 5)	Thailand	OR-R TB Equity	COFCO BIOTECHNOLOGY CL LT-A (DMU 55)	China	000,930 C2 Equity
BANGCHAK CORP PCL-FOREIGN (DMU6)	Thailand	BCP/F TB Equity	CHINA PETROLEUM & CHEMICAL-A (DMU 56)	China	600,028 C1 Equity
BANGCHAK CORP PCL (DMU7)	Thailand	BCP TB Equity	CHINA PETROLEUM & CHEMICAL-A (DMU 57)	China	600,028 CH Equity
BANGCHAK CORP PCL-NVDR (DMU8)	Thailand	BCP-R TB Equity	HENGYUAN REFINING CO BHD (DMU 58)	Malaysia	HYR MK Equity
NORINCO INTL COOPERATION -A (DMU9)	China	000,065 CH Equity	JINKO SOLAR CO LTD-A (DMU 59)	China	688,223 C1 Equity
SICHUAN NEW ENERGY POWER C-A (DMU10)	China	000,155 CH Equity	JINKO SOLAR CO LTD-A (DMU 60)	China	688,223 CH Equity
BCPG PCL-NVDR (DMU 11)	Thailand	BCPG-R TB Equity	ZHEJIANG CHINT ELECTRICS-A (DMU 61)	China	601,877 C1 Equity
BCPG PCL (DMU12)	Thailand	BCPG TB Equity	ZHEJIANG CHINT ELECTRICS-A (DMU 62)	China	601,877 CH Equity

(continued)

Table A (continued)

Entity name	Location	Exchange: Ticker	Entity name	Location	Exchange: ticker
SICHUAN NEW ENERGY POWER C-A (DMU 13)	China	000,155 C2 Equity	SHANGHAI AIKO SOLAR ENERGY-A (DMU 63)	China	600,732 CH Equity
NORINCO INTL COOPERATION -A (DMU 14)	China	000,065 C2 Equity	SHANGHAI AIKO SOLAR ENERGY-A (DMU 64)	China	600,732 C1 Equity
BCPG PCL-FOREIGN (DMU 15)	Thailand	BCPG/F TB Equity	PETRONAS DAGANGAN BHD (DMU 65)	Malaysia	PETD MK Equity
GCL SYSTEM INTEGRATION TEC-A (DMU 16)	China	002,506 C2 Equity	CHINA OILFIELD SERVICES-A (DMU 66)	China	601,808 CH Equity
GCL SYSTEM INTEGRATION TEC-A (DMU 17)	China	002,506 CH Equity	CHINA OILFIELD SERVICES-A (DMU 67)	China	601,808 C1 Equity
PTT EXPLOR & PROD PCL-NVDR (DMU 18)	Thailand	PTTEP-R TB Equity	MALAYSIA MARINE AND HEAVY EN (DMU 68)	Malaysia	MMHE MK Equity
PTT EXPLOR & PROD PUBLIC CO (DMU 19)	Thailand	PTTEP TB Equity	IRPC PCL - NVDR (DMU 69)	Thailand	IRPC-R TB Equity
PTT EXPLORATION & PROD-FOR (DMU 20)	Thailand	PTTEP/F TB Equity	IRPC PCL – FOREIGN (DMU 70)	Thailand	IRPC/F TB Equity
CONOCOPHILLIPS (DMU 21)	Indonesia, China, Vietnam	COP_KZ KZ Equity	IRPC PCL (DMU 71)	Thailand	IRPC TB Equity
SANTOS LTD (DMU 22)	Papua New Guinea	STO PB Equity	XINJIANG DAQO NEW ENERGY C-A (DMU 72)	China	688,303 C1 Equity
OCCIDENTAL PETROLEUM CORP (DMU 23)	Colombia, Pakistan, Oman, UAE	OXY KZ Equity	XINJIANG DAQO NEW ENERGY C-A (DMU 73)	China	688,303 CH Equity

(continued)

Table A (continued)

Entity name	Location	Exchange: Ticker	Entity name	Location	Exchange: ticker
JA-SOLAR TECHNOLOGY CO LTD-A (DMU 24)	China	002,459 CH Equity	PETROCHINA CO LTD-A (DMU 74)	China	601,857 CH Equity
JA-SOLAR TECHNOLOGY CO LTD-A (DMU 25)	China	002,459 C2 Equity	PETROCHINA CO LTD-A (DMU 75)	China	601,857 C1 Equity
TRINA SOLAR CO LTD-A (DMU 26)	China	688,599 C1 Equity	BUMI ARMADA BERHAD (DMU 76)	Malaysia	BAB MK Equity
TRINA SOLAR CO LTD-A (DMU 27)	China	688,599 CH Equity	SINOPEC OILFIELD SERVICE C-A (DMU 77)	China	600,871 CH Equity
ORIENTAL ENERGY CO LTD -A (DMU 28)	China	002,221 C2 Equity	SINOPEC OILFIELD SERVICE C-A (DMU 78)	China	600,871 C1 Equity
ORIENTAL ENERGY CO LTD -A (DMU 29)	China	002,221 CH Equity	SAN MIGUEL CORP (DMU 79)	Philippines	SMC PM Equity
THAI OIL PCL-FOREIGN (DMU 30)	Thailand	TOP/F TB Equity	FLAT GLASS GROUP CO LTD-A (DMU 80)	China	601,865 C1 Equity
THAI OIL PCL (DMU 31)	Thailand	TOP TB Equity	FLAT GLASS GROUP CO LTD-A (DMU 81)	China	601,865 CH Equity
THAI OIL PCL-NVDR (DMU 32)	Thailand	TOP-R TB Equity	NOSTRUM OIL & GAS PLC (DMU82)	Kazakhstan	GB_NTRM KZ Equity
SUZHOU MAXWELL TECHNOLOGIE-A (DMU 33)	China	300,751 CH Equity	CNOOC LTD-A (DMU83)	China	600,938 CH Equity
SUZHOU MAXWELL TECHNOLOGIE-A (DMU 34)	China	300,751 C2 Equity	CNOOC LTD-A (DMU 84)	China	600,938 C1 Equity

(continued)

Table A (continued)

Entity name	Location	Exchange: Ticker	Entity name	Location	Exchange: ticker
HENGDIAN GROUP DMEGC -A (DMU 35)	China	002,056 CH Equity	MING YANG SMART ENERGY GRO-A (DMU 85)	China	601,615 C1 Equity
HENGDIAN GROUP DMEGC -A (DMU 36)	China	002,056 C2 Equity	MING YANG SMART ENERGY GRO-A (DMU 86)	China	601,615 CH Equity
EXXON MOBIL CORP (DMU 37)	Papua New Guinea, Indonesia	XOM_KZ KZ Equity	GOLDWIND SCIENCE & TECHNOL-A (DMU 87)	China	002,202 C2 Equity
PTG ENERGY PCL-NVDR (DMU 38)	Thailand	PTG-R TB Equity	GOLDWIND SCIENCE & TECHNOL-A9 (DMU 88)	China	002,202 CH Equity
PTG ENERGY PCL (DMU 39)	Thailand	PTG TB Equity	PETRON CORP (DMU 89)	Philippines	PCOR PM Equity
PTG ENERGY PCL-FOREIGN (DMU 40)	Thailand	PTG/F TB Equity	SHANGHAI AEROSPACE AUTOMOB-A (DMU 90)	China	600,151 C1 Equity
PETRONAS GAS BHD (DMU 41)	Malaysia	PTG MK Equity	SHANGHAI AEROSPACE AUTOMOB-A (DMU 91)	China	600,151 CH Equity
LONGI-GREEN ENERGY TECHNOL-A (DMU 42)	China	601,012 C1 Equity	PRIMA MARINE PCL-NVDR (DMU 92)	Thailand	PRM-R TB Equity
LONGI-GREEN ENERGY TECHNOL-A (DMU 43)	China	601,012 CH Equity	PRIMA MARINE PCL (DMU 93)	Thailand	PRM TB Equity
PTT PCL/FOREIGN (DMU 44)	Thailand	PTT/F TB Equity	PRIMA MARINE PCL-FOREIGN (DMU 94)	Thailand	PRM/F TB Equity
PTT PCL (DMU 45)	Thailand	PTT TB Equity	SINOPEC OILFIELD EQUIPMENT-A (DMU 95)	China	000,852 C2 Equity
PTT PCL-NVDRN (DMU 46)	Thailand	PTT-R TB Equity	SINOPEC OILFIELD EQUIPMENT-A (DMU 96)	China	000,852 CH Equity
TCL ZHONGHUAN RENEWABLE EN-A (DMU 47)	China	002,129 C2 Equity	HOYUAN GREEN ENERGY CO LTD-A (DMU 97)	Taiwan	603,185 C1 Equity

(continued)

Table A (continued)

Entity name	Location	Exchange: Ticker	Entity name	Location	Exchange: ticker
TCL ZHONGHUAN RENEWABLE EN-A (DMU 48)	China	002,129 CH Equity	HOYUAN GREEN ENERGY CO LTD-A (DMU 98)	Taiwan	603,185 CH Equity
TONGWEI CO LTD-A (DMU 49)	China	600,438 CH Equity	CNOOC ENERGY TECHNOLOGY & -A (DMU 99)	Taiwan	600,968 CH Equity
TONGWEI CO LTD-A (DMU 50)	China	600,438 C1 Equity	CNOOC ENERGY TECHNOLOGY & -A (DMU 100)	Taiwan	600,968 C1 Equity

References

Abramowski, J., & Posorski, R. (2000). Wind energy for developing countries. *DEWI-Magazin, 9*.

Alam, A., Banna, H., Alam, A. W., Bhuiyan, Md. B. U., & Mokhtar, N. B. (2024). Climate change and geopolitical conflicts: The role of ESG readiness. *Journal of Environmental Management, 353*, Article 120284. https://doi.org/10.1016/j.jenvman.2024.120284

Albuquerque, R., Koskinen, Y., & Zhang, C. (2019). Corporate social responsibility and firm risk: Theory and empirical evidence. *Management Science, 65*(10), 4451–4469. https://doi.org/10.1287/mnsc.2018.3043

Al-Refaie, A., Hammad, M., & Li, M.-H. (2016). DEA window analysis and Malmquist index to assess energy efficiency and productivity in Jordanian industrial sector. *Energy Efficiency, 9*(6), 1299–1313. https://doi.org/10.1007/s12053-016-9424-0

Behl, A., Kumari, P. S. R., Makhija, H., & Sharma, D. (2022). Exploring the relationship of ESG score and firm value using cross-lagged panel analyses: Case of the Indian energy sector. *Annals of Operations Research, 313*(1), 231–256. https://doi.org/10.1007/s10479-021-04189-8

Bergeri, I., Whelan, M. G., Ware, H., Subissi, L., Nardone, A., Lewis, H. C., Li, Z., Ma, X., Valenciano, M., Cheng, B., Al Ariqi, L., Rashidian, A., Okeibunor, J., Azim, T., Wijesinghe, P., Le, L.-V., Vaughan, A., Pebody, R., Vicari, A., Yan, T., Yanes-Lane, M., Cao, C., Clifton, D. A., Cheng, M. P., Papenburg, J., Buckeridge, D., Bobrovitz, N., Arora, R. K., Van Kerkhove, M. D. (2022). Global SARS-CoV-2 seroprevalence from January 2020 to April 2022: A systematic review and meta-analysis of standardized population-based studies. *PLoS Medicine, 19*(11), e1004107. https://doi.org/10.1371/journal.pmed.1004107

Buonomano, A., Barone, G., & Forzano, C. (2023). Latest advancements and challenges of technologies and methods for accelerating the sustainable energy transition. *Energy Reports, 9*, 3343–3355. https://doi.org/10.1016/j.egyr.2023.02.015

Chachuli, F. S. M., Mat, S., Ludin, N. A., & Sopian, K. (2021). Performance evaluation of renewable energy R&D activities in Malaysia. *Renewable Energy, 163*, 544–560. https://doi.org/10.1016/j.renene.2020.08.160

Chang, Y.-J., & Lee, B.-H. (2022). The impact of ESG activities on firm value: Multi-level analysis of industrial characteristics. *Sustainability, 14*(21), 14444. https://doi.org/10.3390/su142114444

Charnes, A., Cooper, W. W., & Rhodes, E. (1978). Measuring the efficiency of decision making units. *European Journal of Operational Research, 2*(6), 429–444. https://doi.org/10.1016/0377-2217(78)90138-8

Chen, H.-M., Kuo, T.-C., & Chen, J.-L. (2022). Impacts on the ESG and financial performances of companies in the manufacturing industry based on the climate change related risks. *Journal of Cleaner Production, 380*, Article 134951. https://doi.org/10.1016/j.jclepro.2022.134951

Chen, S., Song, Y., & Gao, P. (2023). Environmental, social, and governance (ESG) performance and financial outcomes: Analyzing the impact of ESG on financial performance. *Journal of Environmental Management, 345*, Article 118829. https://doi.org/10.1016/j.jenvman.2023.118829

Chishti, M. Z., & Sinha, A. (2022). Do the shocks in technological and financial innovation influence the environmental quality? Evidence from BRICS economies. *Technology in Society, 68*, Article 101828. https://doi.org/10.1016/j.techsoc.2021.101828

Cojoianu, T. F., Ascui, F., Clark, G. L., Hoepner, A. G. F., & Wójcik, D. (2021). Does the fossil fuel divestment movement impact new oil and gas fundraising? *Journal of Economic Geography, 21*(1), 141–164. https://doi.org/10.1093/jeg/lbaa027

Damodaran, A. (2007). Return on Capital (ROC), Return on Invested Capital (ROIC) and Return on Equity (ROE): Measurement and Implications. *SSRN Electronic Journal*. https://doi.org/10.2139/ssrn.1105499

Dellnitz, A., Kleine, A., & Rödder, W. (2018). CCR or BCC: What if we are in the wrong model?. *Journal of Business Economics, 88*(7), 831–850. https://doi.org/10.1007/s11573-018-0906-8

Derwall, J., Guenster, N., Bauer, R., & Koedijk, K. (2005). The eco-efficiency premium puzzle. *Financial Analysts Journal, 61*(2), 51–63. https://doi.org/10.2469/faj.v61.n2.2716

Dhar, B. K., Harymawan, I., & Sarkar, S. M. (2022). Impact of corporate social responsibility on financial expert CEOs ' turnover in heavily polluting companies in Bangladesh. *Corporate Social Responsibility and Environmental Management, 29*(3), 701–711. https://doi.org/10.1002/csr.2230

Drakos, A. A., & Bekiris, F. V. (2010). Corporate performance, managerial ownership and endogeneity: A simultaneous equations analysis for the Athens stock exchange. *Research in International Business and Finance, 24*(1), 24–38. https://doi.org/10.1016/j.ribaf.2009.01.002

Ediger, V. Ş., & Berk, I. (2018). Geostrategic Challenges in the Oil and Gas Sectors. In A. B. Dorsman, V. Ş. Ediger, & M. B. Karan (Eds.), *Energy Economy, Finance and Geostrategy* (pp. 173–197). Springer International Publishing. https://doi.org/10.1007/978-3-319-76867-0_9

Egorova, A. A., Grishunin, S. V., & Karminsky, A. M. (2022). The Impact of ESG factors on the performance of Information Technology Companies. *Procedia Computer Science, 199*, 339–345. https://doi.org/10.1016/j.procs.2022.01.041

Freeman, R. E. (2010). *Strategic management: A stakeholder approach.* Cambridge University Press.

Friede, G., Busch, T., & Bassen, A. (2015). ESG and financial performance: Aggregated evidence from more than 2000 empirical studies. *Journal of Sustainable Finance & Investment, 5*(4), 210–233. https://doi.org/10.1080/20430795.2015.1118917

Garcia, A. S., Mendes-Da-Silva, W., & Orsato, R. J. (2017). Sensitive industries produce better ESG performance: Evidence from emerging markets. *Journal of Cleaner Production, 150*, 135–147. https://doi.org/10.1016/j.jclepro.2017.02.180

Gollakota, A. R. K., & Shu, C.-M. (2023). Covid-19 and energy sector: Unique opportunity for switching to clean energy. *Gondwana Research, 114*, 93–116. https://doi.org/10.1016/j.gr.2022.01.014

Grassmann, M. (2021). The relationship between corporate social responsibility expenditures and firm value: The moderating role of integrated reporting. *Journal of Cleaner Production, 285*, 124840. https://doi.org/10.1016/j.jclepro.2020.124840

Hoang, T. (2018). The role of the integrated reporting in raising awareness of environmental, social and corporate governance (ESG) performance. In S. Seifi & D. Crowther (Eds.), *Developments in corporate governance and responsibility* (Vol. 14, pp. 47–69). Emerald Publishing Limited. https://doi.org/10.1108/S2043-052320180000014003

Iazzolino, G., Bruni, M. E., Veltri, S., Morea, D., & Baldissarro, G. (2023). The impact of ESG factors on financial efficiency: An empirical analysis for the selection of sustainable firm portfolios. *Corporate Social Responsibility and Environmental Management, 30*(4), 1917–1927. https://doi.org/10.1002/csr.2463

Jiang, W., & Martek, I. (2021). Political risk analysis of foreign direct investment into the energy sector of developing countries. *Journal of Cleaner Production, 302*, Article 127023. https://doi.org/10.1016/j.jclepro.2021.127023

Khalil, R. G., Damrah, S., Bajaher, M., & Shawtari, F. A. (2023). RETRACTED ARTICLE: Unveiling the relationship of ESG, fintech, green finance, innovation and sustainability: Case of Gulf countries. *Environmental Science and Pollution Research, 30*(54), 116299–116312. https://doi.org/10.1007/s11356-023-30584-8

Kim, S., & Li, Z. (Frank). (2021). Understanding the Impact of ESG Practices in Corporate Finance. *Sustainability, 13*(7), 3746. https://doi.org/10.3390/su13073746

Kuzemko, C., Bradshaw, M., Bridge, G., Goldthau, A., Jewell, J., Overland, I., Scholten, D., Van De Graaf, T., & Westphal, K. (2020). Covid-19 and the politics of sustainable energy transitions. *Energy Research & Social Science, 68*, Article 101685. https://doi.org/10.1016/j.erss.2020.101685

Li, T.-T., Wang, K., Sueyoshi, T., & Wang, D. D. (2021). ESG: Research progress and future prospects. *Sustainability, 13*(21), 11663. https://doi.org/10.3390/su132111663

Liu, P., Zhu, B., Yang, M., & Chu, X. (2022). ESG and financial performance: A qualitative comparative analysis in China's new energy companies. *Journal of Cleaner Production, 379*, Article 134721. https://doi.org/10.1016/j.jclepro.2022.134721

Miralles-Quirós, M., Miralles-Quirós, J., & Valente Gonçalves, L. (2018). The value relevance of environmental, social, and governance performance: The Brazilian case. *Sustainability, 10*(3), 574. https://doi.org/10.3390/su10030574

Mishra, S., & Modi, S. B. (2013). Positive and negative corporate social responsibility, financial leverage, and idiosyncratic risk. *Journal of Business Ethics, 117*(2), 431–448. https://doi.org/10.1007/s10551-012-1526-9

Morozov, I. V., Potanina, Y. M., Voronin, S. A., Kuchkovskaya, N. V., & Mapo, D. S. (2018). Prospects for the development of the oil and gas industry in the regional and global economy. *International Journal of Energy Economics and Policy, 8*(4), 55–62. Retrieved from https://www.proquest.com/scholarly-journals/prospects-development-oil-gas-industry-regional/docview/2083002289/se-2

Naeem, N., & Çankaya, S. (n.d.). *The impact of ESG performance over financial performance: A study on global energy and power generation companies.*

Nirino, N., Santoro, G., Miglietta, N., & Quaglia, R. (2021). Corporate controversies and company's financial performance: Exploring the moderating role of ESG practices. *Technological Forecasting and Social Change, 162*, Article 120341. https://doi.org/10.1016/j.techfore.2020.120341

Orlitzky, M., Schmidt, F. L., & Rynes, S. L. (2003). Corporate Social and financial performance: A meta-analysis. *Organization Studies, 24*(3), 403–441. https://doi.org/10.1177/0170840603024003910

Painuly, J. P., & Wohlgemuth, N. (2021). Renewable energy technologies: Barriers and policy implications. In *Renewable-energy-driven future* (pp. 539–562). Elsevier.

Petersen, M. A. (2009). Estimating standard errors in finance panel data sets: Comparing approaches. *Review of Financial Studies, 22*(1), 435–480. https://doi.org/10.1093/rfs/hhn053

Popkova, E. G., & Sergi, B. S. (2021). Energy efficiency in leading emerging and developed countries. *Energy, 221*, 119730. https://doi.org/10.1016/j.energy.2020.119730

Ruan, L., & Liu, H. (2021). Environmental, social, governance activities and firm performance: Evidence from China. *Sustainability, 13*(2), 767. https://doi.org/10.3390/su13020767

Suárez-Fernández, Ó., Maside-Sanfiz, J. M., López-Penabad, M. C., & Alzghoul, M. O. (2024). Do diversity & inclusion of human capital affect ecoefficiency? Evidence for the energy sector. *Green Finance, 6*(3), 430–456. https://doi.org/10.3934/GF.2024017

Shahsavari, A., & Akbari, M. (2018). Potential of solar energy in developing countries for reducing energy-related emissions. *Renewable and Sustainable Energy Reviews, 90*, 275–291. https://doi.org/10.1016/j.rser.2018.03.065

Siakwah, P., Musavengane, R., & Leonard, L. (2020). Tourism governance and attainment of the sustainable development goals in Africa. *Tourism Planning & Development, 17*(4), 355–383. https://doi.org/10.1080/21568316.2019.1600160

Siksnelyte-Butkiene, I. (2021). Impact of the COVID-19 pandemic to the sustainability of the energy sector. *Sustainability, 13*(23), 12973. https://doi.org/10.3390/su132312973

Singh, K. M., Meena, M. S., Kumar, A., & Singh, R. K. P. (2011). *Socio-economic determinants of rural poverty: An empirical exploration of Jharkhand state.* SSRN Electronic Journal. https://doi.org/10.2139/ssrn.2017593

Singhania, M., Saini, N., Shri, C., & Bhatia, S. (2024). Cross-country comparative trend analysis in ESG regulatory framework across developed and developing nations. *Management of Environmental Quality: An International Journal, 35*(1), 61–100. https://doi.org/10.1108/MEQ-02-2023-0056

Sinha, A., Sengupta, T., & Alvarado, R. (2020). Interplay between technological innovation and environmental quality: Formulating the SDG policies for next 11 economies. *Journal of Cleaner Production, 242*, Article 118549. https://doi.org/10.1016/j.jclepro.2019.118549

Van Der Linden, B., & Freeman, R. E. (2017). Profit and other values: Thick evaluation in decision making. *Business Ethics Quarterly, 27*(3), 353–379. https://doi.org/10.1017/beq.2017.1

Vázquez-Rowe, I., & Iribarren, D. (2015). Review of life-cycle approaches coupled with data envelopment analysis: Launching the CFP + DEA method for energy policy making. *The Scientific World Journal, 2015*, 1–10. https://doi.org/10.1155/2015/813921

Wang, Q., Huang, R., & Li, R. (2022). Towards smart energy systems—A survey about the impact of COVID-19 pandemic on renewable energy research. *Energy Strategy Reviews, 41*, Article 100845. https://doi.org/10.1016/j.esr.2022.100845

Woo, C., Chung, Y., Chun, D., Seo, H., & Hong, S. (2015). The static and dynamic environmental efficiency of renewable energy: A Malmquist index analysis of OECD countries. *Renewable and Sustainable Energy Reviews, 47*, 367–376. https://doi.org/10.1016/j.rser.2015.03.070

Yatim, P., Mamat, M.-N., Mohamad-Zailani, S. H., & Ramlee, S. (2016). Energy policy shifts towards sustainable energy future for Malaysia. *Clean Technologies and Environmental Policy, 18*(6), 1685–1695. https://doi.org/10.1007/s10098-016-1151-x

Yoon, B., Lee, J. H., & Byun, R. (2018). Does ESG performance enhance firm value? Evidence from Korea. *Sustainability, 10*(10), 3635. https://doi.org/10.3390/su10103635

Zhan, J. X., & Santos-Paulino, A. U. (2021). Investing in the sustainable development goals: Mobilization, tg, and impact. *Journal of International Business Policy, 4*(1), 166–183. https://doi.org/10.1057/s42214-020-00093-3

Zhang, Y., Li, X., Jiang, F., Song, Y., & Xu, M. (2020). Industrial policy, energy and environment efficiency: Evidence from Chinese firm-level data. *Journal of Environmental Management, 260*, Article 110123. https://doi.org/10.1016/j.jenvman.2020.110123

Zhou, G., Liu, L., & Luo, S. (2022). Sustainable development, ESG performance and company market value: Mediating effect of financial performance. *Business Strategy and the Environment, 31*(7), 3371–3387. https://doi.org/10.1002/bse.3089

MIX
Papier aus verantwortungsvollen Quellen
Paper from responsible sources
FSC® C105338

FSC
www.fsc.org

If you have any concerns about our products,
you can contact us on
ProductSafety@springernature.com

In case Publisher is established outside the EU,
the EU authorized representative is:
Springer Nature Customer Service Center GmbH
Europaplatz 3, 69115 Heidelberg, Germany

Printed by Libri Plureos GmbH
in Hamburg, Germany